Environmental Stewardship

Environmental Stewardship

J. DOUMA

EDITED BY NELSON D. KLOOSTERMAN
TRANSLATED BY ALBERT H. OOSTERHOFF

WIPF & STOCK · Eugene, Oregon

ENVIRONMENTAL STEWARDSHIP

Originally published as *Milieu en manipulatie*, 2nd edition, by J. Douma (Kampen: Van den Berg, 1988).

Copyright © 2015 The Publication Foundation of the Theological College of the Canadian Reformed Churches. All rights reserved. Except for brief quotations in critical publications or reviews, no part of this book may be reproduced in any manner without prior written permission from the publisher. Write: Permissions, Wipf and Stock Publishers, 199 W. 8th Ave., Suite 3, Eugene, OR 97401.

Scripture quotations taken from The Holy Bible, English Standard Version®, copyright © 2001 by Crossway, a publishing ministry of Good News Publishers. Used by permission. All rights reserved.

Wipf & Stock
An imprint of Wipf and Stock Publishers
199 W. 8th Ave., Suite 3
Eugene, OR 97401

www.wipfandstock.com

isbn 13: 978-1-4982-0600-6

Manufactured in the U.S.A.

Contents

Editor's Introduction | *vii*

1 Environment: Damage and Guilt | 1
2 The Bible and the Environment | 25
3 The Agenda: What Needs to Be Done? | 48
4 Genetic Engineering (1): Plants and Animals | 72
5 Genetic Engineering (2): Human Beings | 93

Bibliography | *123*
Scripture Index | *129*
Subject Index | *133*

Editor's Introduction

Professor J. Douma (pronounced *Dow-ma*) served for most of his life as a minister among the Reformed Churches in the Netherlands, and since 1970 has been Professor of Ethics (now emeritus) at their Theological University in Kampen. Doctor Douma has written, among other works, a fifteen-volume series, entitled *Ethical Reflection* (*Ethische bezinning*), in which he discusses various fundamental and up-to-date subjects in the area of ethics. These subjects include abortion, marriage and sexuality, Christian lifestyle, homosexuality, environment and technology, political responsibility, and nuclear armament.

Doctor Douma is respected internationally for his thoughtful interpretation and careful application of Scripture, the church's creeds, and church history in relation to contemporary moral problems. In this volume he provides an analysis of environmental stewardship and of the special calling of Christ-followers to practice such stewardship of the creation they joyfully confess to be God's gift to humanity.

Please note these acknowledgements of gratitude. We express our hearty thanks to Dr. Albert H. Oosterhoff for his attentive labors in translating this work, and to Ms. Deanna Smid for her capable work in formatting and proofreading the manuscript. Laboring together in partnership on The Douma Project, the Canadian Reformed Theological Seminary and Worldview Resources International express special thanks to Mr. Al Schutten and Mr.

Editor's Introduction

Barry Hordyk, who have assisted in funding this volume of The Douma Project.

Nelson D. Kloosterman
Director, The Douma Project
Worldview Resources International
St. John, Indiana, USA
October 13, 2014

1

Environment
Damage and Guilt

A SAD DEVELOPMENT

ESPECIALLY SINCE 1970 WE have become aware of the destruction that human beings can wreak on the environment. The report to the Club of Rome published in 1972 (titled *The Limits to Growth*) created a sensation. Using computer models, researchers from the Massachusetts Institute of Technology calculated that the limits of population and economic growth would be reached within a hundred years. Moreover, the report detailed, the limits of environmental pollution would be reached within the foreseeable future.

The researchers at MIT explained how five factors (population, food production, industrialization, pollution, and the use of non-replaceable natural resources) influence each other and thereby grow *exponentially*. As an example, a child who grows half an inch taller each year grows linearly, but a sum of money invested at 7 percent interest grows exponentially. Each year the interest increases the capital, so that the next year the sum of money has increased by a specific exponent. That, of course, is good for the

1

investor. But there are also damaging forms of exponential growth, with serious consequences for the environment. The report to the Club of Rome compared the growth of human society to the pattern of a water lily growing in a pond. The lily doubles in size every day, so if it remains undisturbed, it will cover the pond in thirty days and choke off all other life in the pond. When the lily looks small, you do not worry yet about cutting it back, until the moment half the pond is covered: "On what day will that happen? On the twenty-ninth day, of course. You have only one day to save your pond."[1]

In 1974, the Club of Rome received a second report revising the methodology of the first report.[2] The first report had described the world as one complete entity and its predictions therefore applied to the whole world. Understandably, countries in the Third World in particular criticized that methodology. Surely predictions about the rich northern hemisphere of the world did not necessarily apply also to the poor southern hemisphere, did they? Besides, the first report essentially concluded with a plea for "zero growth," and developing countries (among others) understandably criticized the report for this reason.

A *regionalized* world model, in which the world was divided into ten regions, replaced the undifferentiated methodology of the first report. Still, the second report communicated the same seriousness as the first. Without preventive action, the consequences of the current trends would be catastrophic fifty years from the date of the report.

In 1988, one of the authors of the second report, Eduard Pestel, delivered a new report to the Club of Rome. In it he analyzed the mistakes of the first report and also rejected the idea of "zero growth." However, he proposed that, rather than carrying on with the current undifferentiated growth, we must instead strive to promote what he called "organic growth and development." By that phrase he meant that *together* we must determine what may or may not be allowed to grow. Currently there is unchecked growth, but

1. Meadows et al., *Limits to Growth*.
2. Mesarovic and Pestel, *Turning Point*.

organic growth demands recognizing mutual interdependence, which does not permit any part of the world to grow at the expense of another part. The mutual interdependence of countries and regions, Pestel argued, is a fact, not a matter of choice.[3] Economist K. E. Boulding has compared the earth to a spaceship: a spaceship is limited unless equipped with an energy source and a food supply. According to Boulding, the earth is limited like a spaceship and if we do not cease our current wasteful "cowboy economy" in favor of a "spaceship economy," the consequences will be dire.[4]

These are alarming messages that have not lost their relevance at all, although there has been widespread criticism of the reports to the Club of Rome. Looking back on the discussion, Pestel said in 1988 that the 1971 report was but a first, defective step. But that step generated a lasting and necessary interest in the future of humankind.

The commotion the report caused (especially in the Netherlands) may have dissipated; however, certain things have become part of public consciousness as a result of the discussions generated by the report: the world is finite and there are *limits* to growth. Indeed, we have a responsibility for the future, a responsibility we cannot escape.

The topic I raise in this chapter poses questions like: What is our attitude and relationship to the environment? We know that it is essential to our existence. Are we destroying it and, if so, how did that happen? What is necessary in order for us to change course so that we interact with the environment in a responsible way?

EXAMPLES OF ENVIRONMENTAL POLLUTION

By the *environment*, we mean our physical, inanimate, and animate surroundings, with which we have a mutual relationship.[5] Often we use the terms *ecosystem* and *ecology* as well. Tellingly, the latter

---

3. Pestel, *Beyond the Limits to Growth*.
4. Boulding, "Economics of the Coming Spaceship Earth," 3–14.
5. Boersema et al., *Basisboek Milieukunde*, 18.

Environmental Stewardship

two terms derive from the Greek word *oikos*, which means "household." Without housing or shelter, people will die. Just like all other living beings, they too depend upon vital life processes that occur between animate elements (such as plants, bacteria, animals, and people) and inanimate elements (such as air, water, minerals, and technological installations). Without such processes, there would be no life on earth. When an ecological problem arises in inanimate matter, many animate species may degenerate or become extinct, and, indeed, human life itself is endangered.

Many ecosystems have already been damaged. Let me give a number of examples. *Deforestation* is an alarming phenomenon throughout the world. People need firewood and timber, agriculture claims a lot of forests, and reforestation occurs rarely. Almost half of the primitive forest has already disappeared, despite the fact that precisely there we encounter an astounding variety of plant and animal species. Rapid deforestation and heavy erosion cause the soil to deteriorate, beginning the process of desertification. The demand for firewood and food remains and grows as the population increases and expands, resulting in continuing deforestation. Exporting wood to wealthy countries causes additional and extensive deforestation. Such rapaciousness seriously affects our climate. Significant droughts and enormous floods are often the result, for deforested areas can no longer retain moisture.

Anyone who thinks that this happens only in tropical regions is mistaken. Damage to forests in the Alps, for example, is having alarming consequences. Tourism is the source of much of the problem. Because of the prosperity of western European countries, summer and winter tourism has increased exponentially. A network of roads, ski runs, parking lots, and other tourist amenities has caused the disappearance of large areas of forest. Many tourists also means many cars. Exhaust fumes have caused acid rain, which in turn has impaired the vitality of alpine forests.

But we can see the effects of deforestation even closer to home. Our Dutch forests, too, are being threatened by acid rain, which happens when pollutants in the air (both in solid form and dissolved in rain, hail, and fog) fall to the ground. Trees and other plants collect most of the pollutants and absorb them into their leaves. The rest falls through runoff into the ground in high concentrations. Then the same trees and plants derive their nutrition from the polluted soil, often with disastrous consequences. The worst pollutants are sulfur dioxide and carbon dioxide, which are released through burning fossil fuels such as coal, oil, and gas. In the Netherlands there is an additional pollutant as well, namely, ammonia, a byproduct of the great quantities of manure generated by the farming industry.

Forests have been called the lungs of the earth. But *water*—seas, rivers, lakes, and ground water—is also vital to our ecosystem. It is widely known that industries impair the quality of surface water by means of chemical and thermal pollution. Illegally dumping oil and poisonous substances pollutes the oceans. Additionally, eutrophication (the addition of excessive nutrients) happens when sewage, pulp from paper mills, and the effluent from slaughterhouses overburdens rivers and lakes with biodegradable material. Regularly we are alarmed by news about large poisonous dumps, which can pollute the groundwater, and consequently also our drinking water, for a certain amount of time.

Air pollution is also a serious matter. I have already drawn attention to gases that cause acid rain and thereby harm forests and the soil. But certain gases that end up in the atmosphere can also endanger the ozone layer, which protects the earth from harmful ultraviolet rays. The use of CFCs (chlorofluorocarbons) as propellants can damage the ozone layer, creating holes in it. When that happens, ultraviolet rays become dangerous for people and animals.

CFCs and other gases also affect the temperature in the atmosphere. The sun heats the surface of the earth and the earth returns infrared rays back into the universe. However, the carbon dioxide in the atmosphere bounces some of those rays back to the earth. If the carbon dioxide in the atmosphere increases, a "greenhouse effect" is created. A denser layer of carbon dioxide causes temperatures to increase, ice sheets to melt, and sea levels to rise, inevitably resulting in catastrophic floods.

Significant damages caused by a complete climate change are likely to occur only in the long term. More immediately dangerous is air pollution, which we experience now already because of our use of energy sources. We have only to think of vehicle exhaust fumes and the discharge from residential and factory chimneys, which cause smog. Moreover, accidents at nuclear power plants, such as those in Harrisburg (1979) and Chernobyl (1986), caused increased radioactivity in the atmosphere. As a precaution, farmers in the Netherlands put their cows back in their barns despite the beautiful spring weather, because the fire at the nuclear generating station in Chernobyl resulted in excessively high levels of radioactivity over large areas of Europe.

As a result of deforestation and pollution of all kinds, many animal and plant species have already become extinct. In addition, intensive hunting and fishing have wiped out entire species of animals. Polluted lakes and seas cause massive loss of marine life. Mercury, dumped into the sea, can be found in the livers of many fish. In her book *Silent Spring*, Ruth Carson writes about the consequences of using pesticides in farming, gardening, and forestry. Nature is comprised of reciprocal relationships, which explains why damage to one of the links endangers the entire chain of the biological process. As the saying goes, "The sparrow-hawk ate the song bird, which ate the beetle, which ate the caterpillar, which ate the leaf of the tree." Therefore, if we want to tackle the enemy, it may happen that a friend will die too.

The so-called *persistent* pesticides cause serious problems. They require an excessively long time to break down in nature (if they do at all). For example, DDT can still be found in the fatty tissues of penguins and seals that live in and around the South Pole.

Various animals are being mistreated in other ways as well. In some sectors of the intensive farming industry, an animal can be regarded as simply a product or a means of production. Producers replace the natural environment of pigs and chickens with a biochemical process designed to manufacture the most desirable product. Calves, for instance, are raised in small, dark pens and fed iron-free food in order to become anemic, thereby ensuring white meat for export. Geese are overfed to produce the enlarged livers desired for the preparation of *foie gras*.

I have given only a few examples from the available evidence in order to demonstrate the extent of our environmental problems.[6] Yet the examples above are sufficient to give an indication of the seriousness of the situation.

THE QUESTION OF GUILT

We often try to assign blame when we realize the extent of the environmental crisis. How could this have happened? Critics point the finger at a large number of guilty parties. Let me mention a few.

Some consider the cause of all the problems to be the *disappearance of animism*. If only the pagan fear of the powers of nature had remained with us, then people would not assault the environment as they do now! Others deplore the influence of the *Judeo-Christian tradition*, in which Genesis 1:28, which speaks about subduing the earth and ruling over all other living creatures, is so significant. Yet others blame the zeal for work so often espoused by *Calvinism*, which is thereby found culpable for the magnitude of environmental pollution.

Philosophers are also named as guilty parties. If only, since Descartes, they had not thought so anthropocentrically about humanity! Descartes regarded the rest of nature merely as an extension of human aspirations, and he thought of animals as machines. Such ideas are closely linked to the belief in progress, an idea that entered the public consciousness particularly through philosophical movements like the Renaissance and the Enlightenment.

Other critics do not assign blame to a specific religion or philosophy; rather, they blame material factors (in addition to ideological ones) that have caused environmental pollution. These factors include overpopulation, capitalism, technology, industrialization, and economic growth.

---

6. For more detailed information, see Reijnders, *Pleidooi voor een duurzame relatie*; and Crul et al., *Natuurlijke hulpbronnen*. A somewhat older source is Voous, *Natuur*, which provides a survey as concise as it is clear.

Environmental Stewardship

SPECIAL GUILTY PARTIES?

I do not think it profitable to identify special groups that can be identified as *the* guilty parties, whether they be Calvinists, Cartesians, or others. My reasons are these.

First, environmental pollution is a phenomenon appearing in every age, not just our own. We have come to recognize it as a worldwide problem in our time, but earlier people also damaged the environment to varying extents. For example, it is a mistake to think that all the deserts have always existed. People turned forests into denuded places, savannahs into steppes, and grasslands into deserts. Libya once had great cities and a productive hinterland; now it is largely a dry, infertile country. As well, the area between Afghanistan and Egypt once supported a rich culture; evidence still remains of its prehistoric irrigation canals, now filled with sand, silted up and salty.[7]

Several species of animals have been extinct for thousands of years, also because of human intervention. Evidence strongly indicates that prehistoric hunters (who were animists!) caused the extinction of several species of large animals. The cultures of Assyria, Babylon, Syria, and Greece all perished because they misused their soil. The Egyptian gold mines were already exhausted during the time of the pharaohs. Slaves were subjected to the hazards of lead when they were forced to work in lead mines and smelting factories. Moreover, noise and water pollution have always accompanied cities. The ancient city of Rome, for example, was notorious for its oppressive and unhealthy air quality.[8]

Second, people often did not realize what the consequences of their actions might be. They simply did not know how particular actions would affect the environment. We should hold people guilty if they convert excessive amounts of forest into farmland when they know the consequences of that practice. But often the consequences were unknown at the time. In the Golden Age, most

---

7. Van Steenis, *Homo destruens*, 13–14.
8. Reijnders, *Pleidooi*, 17–18; Voous, *Natuur*, 22, 38.

of our forests were cut down. Hol-land[9] was once a wood-land. People cut down the forests to obtain arable land without giving thought to necessary reforestation. When serious floods occurred in the Alps in the eighteenth century, people did not know then what we know now. Only by about the year 1800 did we discover that floods and landslides have something to do with deforestation.

But today as well we do not always correctly anticipate the consequences of our technological actions. The construction of the Aswan Dam on the Nile in the 1960s is an example of this. The dam was built to generate electricity and to improve the water supply. But the new lake behind the dam silted up with sand carried downstream by the Nile. This turned the lake into a breeding ground for insects, which caused the spread of diseases. Moreover, the fertile silt no longer reached the downstream areas, which in turn lost their fruitfulness. Many fish downstream from the dam also disappeared because the Nile no longer delivered their food supply and, in consequence, the local population lost an important income source. Downstream, agricultural lands along the edges of the river silted up because there were no longer regularly inundated.[10] Thus, much can go wrong as a result of one action, and therefore we should not immediately accuse others of exploitation and egoism. Often we simply do not know all the possible harmful consequences of even well-intentioned plans.

Third, even if people are not aware of the consequences—and who today is not?—we should be wary of condemning others. It is easy to condemn all excessive cultivation and other types of environmental pollution while we enjoy a reasonable or very good lifestyle ourselves. But what about the poor person in Africa or South America who is struggling to secure the bare necessities of life? Forests in the tropics and elsewhere were a great obstacle to people, and in their struggle for existence they had to overcome them first.[11] Many people are still engaged in a similar struggle. It

---

9. *Translator's note*: The word "Holland" is hyphenated in the original to parallel the word "wood-land" and is intended as a pun. "Hol" means "empty."

10. The example was given by Bibo, "Wat is er aan de hand," 24.

11. Van Steenis, *Homo destruens*, 7.

is easy for us to speak of the "lush tropical vegetation" that must not be disturbed, but that term, while *botanically* correct, obscures the basic necessities that this vegetation offers people in an early stage of technological development.[12]

Also, for many in the industrialized world, the call to keep the environment safe sounds cheap when they work in industries that cause a lot of pollution. The environment may be a problem, but so are poverty and striving to provide for the necessities of life. Those who condemn the destruction of the environment should also point out alternatives. Otherwise, they are choosing trees, animals, and beautiful rivers over human beings.

The environmental problem is a *structural* one; it is concerned with more than pollution. When we understand the full extent of the interrelationship between the environment and the economy, we realize that we are dealing with a complex problem. If one person endangers the environment because of poverty, the danger actually may be caused by the wealth of another person who behaves like Cain, who asked, "Am I my brother's keeper?" This should make us careful when looking for guilty parties, as we are not necessarily innocent.

One thing remains the same in every search for the guilty party: it is always people who are at fault. Humanity's fall into sin is also the cause of the environmental crisis. People and the environment stand against each other as enemies. In the difficult struggle to survive, people will find a way to overcome obstacles. Their ignorance of the consequences of their actions, and their shortsightedness and unbridled egoism are responsible for much harm, until their eyes are opened and they see the worldwide damage. Even before 1970, books warned about "our pillaged planet," the "rape of the earth," and "the earth that is dying."[13]

People are now sounding the alarm out of self-interest. We shall see that we need more than an understanding of our

---

12. Ibid.

13. The quoted phrases are titles used in Slijper, *Het lot*, 8, where the author himself notes, "Human beings have, indeed, acted like animals on the earth."

self-interest to gauge the extent of the environmental crisis and to avert it.

CHRISTIANITY AND THE ENVIRONMENT

I noted above that Christianity has also been accused as the cause of environmental destruction. Because this accusation is based upon certain interpretations of important scriptural passages, I need to pay closer attention to it.

According to Claus Jacobi, people who belong to other races and faiths have made the same mistakes as Christians, but they did so out of ignorance. No one has ruined nature on such a large scale, however, and in such a self-willed way, as Christians, he posits. Taoism demands in its seventh commandment, "You shall not burn off the grassland, or set fire to forests on mountains," but the Bible's Ten Commandments are concerned exclusively with human beings, according to Jacobi. The Christian God commands man to rule over the fish of the sea, the birds, and all other living creatures, and to subdue the earth. From that point of view, man is not part of nature; rather, he is its master, an attitude and position that caused the destruction of animal species and the pollution of the soil.[14]

Jacobi's facile conclusions have been adopted without argument by many people. By contrast, Lynn White's position is much more nuanced. This historian and student of the Middle Ages published an article in 1967 that attracted much attention. White argued that in the ninth century (before the environmental crisis) people's relationship to the earth changed drastically because of the invention of a new plow, shortly before the year 830. This plow went deeper into the ground, and so pulling it required eight oxen rather than two. Because it was uncommon for one farmer to own eight oxen, cooperation with other farmers became necessary. Such cooperation in turn led to a more efficient use of the soil.

---

14. Jacobi, *De menselijke springvloed*, 32–33.

What was the result? Whereas before then, people were *part* of nature, thereafter they became *exploiters* of nature. People and nature became two separate entities, and people were the master.

Although others also regard technological developments as the prime cause of environmental damage and identify Copernicus, Descartes, or (later) the Industrial Revolution as the guilty parties, White goes back further. He cites important discoveries that predate Descartes and others, such as the watermill, the windmill, and the mechanical clock, which had been in use for a long time already. The technological superiority of European powers at the end of the fifteenth century was already so great that they were able to conquer, plunder, and colonize the rest of the world.

White believes that man's relationship with the environment is strongly influenced by religion. That is true not only in India or Sri Lanka, but also in Christian countries. According to White, we must regard the victory of Christianity over paganism as the greatest psychical revolution in the history of our culture. Pagans, such as Aristotle, believed in a nature that had no beginning and that would be renewed eternally, but Christianity taught a creation that had a purpose. God made all kinds of creatures on earth for the benefit of humanity, to whom everything was to be subservient. While the human body was formed from the dust of the earth, in God's image, people are not merely a part of nature. Especially in its Western form, Christianity is the most anthropocentric religion that the world has ever seen, according to White.

In order to define the contrast between paganism and Christianity more clearly, White points out that for pagans every tree, every spring, every stream, and every hill has its own guardian spirit. No tree could be cut down, mineshaft dug, or stream dammed before the relevant spirit was placated. By destroying pagan animism, Christianity thereby made exploitation of nature possible. Thereafter, people were indifferent to the "feelings" of natural objects. While it is true that Christian saints replaced pagan spirits, that did not guarantee the protection of nature. Christian saints, after all, did not reside *in* natural objects, as was the case with pagan spirits. Rather, their citizenship is in heaven.

Environment

If, together with White, we assume that the source of Western technology (and along with that, all environmental misery) resides in the ninth century, we may wonder why technology such as we now enjoy took so long to develop. White's answer is that technology and science were independent of each other for centuries because, according to him, they were practiced by different classes of people. The study of *science* was an aristocratic, speculative, and intellectual exercise, whereas *technology* developed among the lower classes, which were concerned with action and manual labor. The democratic revolutions in the nineteenth century removed those distinctions, and as a result we ended up with a functional unity of theory and practice. Only in that democratic climate did the marriage of science and technology become possible. Therefore, according to White, our ecological crisis is the product of democratic culture.[15]

CALVINISM

I have given an extensive summary of White's argument because in his article he echoes a commonly held belief: the environmental crisis is a result of Western technology, Western technology is rooted in Christianity, and Christianity in turn derives its vision of nature from the Bible. Thus, whether it is understood in simplistic or nuanced terms, the Bible is held responsible for much of the current environmental pollution because its story of the creation of the world assigns humanity a special place within that world.

Many variations on this theme of blame exist. Here is one: sometimes Protestantism, and more specifically, *Calvinism*, are identified as special perpetrators. The industriousness of Protestants in business and industry supposedly generated the ugly consequences that accompanied the triumphal march of technology. Max Weber (1864–1920) saw a connection between the Calvinist doctrine of election and the rise of capitalism. The doctrine of election supposedly created great uncertainty for many people. How

15. White, "Historical Roots," 1203–7.

can I know that I am chosen? People wanted certainty and looked for it in external signs, like industry and wealth. This desire for assurance caused Calvinists to work hard. Their austere lifestyle, demanded by God, also ensured that they amassed a lot of money. The "innerweltliche Askese" [worldly asceticism] of the Calvinistic Puritans, according to Max Weber, led to accumulating capital and, thus, to capitalism, with all the adverse consequences thereof!

Of course, Max Weber, who died in 1920, did not yet consider our environmental crisis. But many use the connection he laid between Calvinism and capitalism to link pollution in a capitalist society with those who cause it.

DOCTRINE AND LIFE ARE TWO DISTINCT THINGS

In addition to decrying the negative influence of the Christian faith, which supposedly caused excessive damage to the environment, many cite contrary examples from other faiths. These citations suggest that if Christianity had never existed we would not be saddled with the current environmental problems, or at any rate things would be much better than they are now. There is every reason to doubt such a thesis.

I have already pointed out that long before the rise of Christianity people caused damage to the environment in a variety of ways. Not only Christian people, but humanity in general is "*Homo destruens*"—a destructive being. Yet it is true that various faiths have made pronouncements that seem to prescribe a gentler attitude toward the environment. Let me mention a few such pronouncements.

Muslims understand from the Koran that, since God ordered the earth, they may not harm it. Polluting water and cutting down trees are forbidden, and camels and horses have to be well cared for.[16]

---

16. M. Tworushcka, in *Umwelt*, 52ff.

Environment

Buddhism calls for people to suffer alongside and to do good to animals and plants as well as other people, for Buddhism regards all forms of life as part of Buddhist enlightenment. "Grass, trees, and the earth—all become Buddha." Indeed, because of their religious conviction, many Buddhists are vegetarians out of respect for animal life.[17]

In Hinduism all life is holy as well, and accordingly people must live in peace and harmony with their fellow creatures. In this view, nature is not an *Umwelt*,[18] which merely forms the outer wings of the stage, while humanity occupies center stage of the drama; rather, nature is the *Mitwelt*[19] of humanity, and humanity forms part of the same life as rivers, trees, and animals.[20]

The Shintoists of Japan and the Native Americans in the United States also exhibit the same universal solidarity with every living thing. Seattle, the chief of the Duwamish (and Suquamish) tribe(s), said in 1885 when the United States wanted to buy the tribal lands, "We are part of the earth and it is part of us. The perfumed flowers are our sisters, the deer, the horse, the great eagle, these are our brothers. The rocky crest, the juices in the meadows, the body heat of the pony, and the man, all belong to the same family."[21]

Despite views like these, their adherents continued to hunt, murder, fell trees, and burn forests. Doctrine and life are two distinct things, just as in Christianity. No one can deny that Christians have damaged the environment, but that does not mean that this was necessarily a result of Christian doctrine. That is true also

17. Hamer and Neu, in ibid., 78ff.

18. *Translator's note:* That is, the "environment," or the world around us. It is a term used in existential philosophy.

19. *Translator's note:* That is, the world of fellow creatures. It too is a term used in existential philosophy.

20. Van Dijk, in *Umwelt*, 119.

21. Quoted in Bouma, *Slagveld*, 16. *Translator's note*: There is apparently considerable doubt whether Chief Seattle delivered the speech in question. See Clark, "Thus Spoke Chief Seattle." Chief Seattle died in 1866. The date of the speech is generally thought to be March 11, 1854, and the 1885 date referred to in the text probably refers to a transcription of the notes of a supposed witness.

of non-Christian faiths. Those faiths may regard all nature as holy, but the facts indicate that not everything was or is treated as holy.

This distinction between doctrine and life is understandable, of course. For it demonstrates not only that people do not adhere to their faith, but also that it is *impossible* to do so completely. You can declare everything to be divine, but you cannot treat everything as divine, for example. As long as *sublime* nature is at the same time a *threatening* nature, the attitude of people toward nature will be ambivalent. In our temperate climate and with our Western conveniences, it is easy for us to say that nature fascinates us, but in earlier times (and for many still) it was a threat[22] that caused people to grab fire, axe, and rifle to defend themselves. For that reason, only *particular* animals, *particular* trees, and *particular* rivers were treated as holy, while others were killed or felled. The cow is not regarded as holy everywhere and was not even always so regarded in India.

We shall see later that a pantheistic vision of nature conflicts with the Bible's revelation. But it should already be clear that non-Christian religions with their pantheistic views only *appear* to offer more guarantees for a healthy environment than the Christian religion. Otherwise, it would be difficult to explain how a non-Christian country like Japan, with its Shinto religion, has been polluted more than most other countries.[23]

It has been noted frequently that different "ecological" ethics do not lead to different results for the environment. In ancient Rome, which stood at the zenith of its political and economic power and which used lofty words to address "mother earth," holy hedges, springs, and high mountains, there simultaneously existed an aggressive attitude toward nature that endangered the soil and animals.[24]

---

22. Klöcker and Tworuschka, *Umwelt*, 176. Today we emphasize the moment of the *fascinans* (that which captivates), whereas in older cultures the *tremendum* (that which causes one to shudder) was always evident.

23. Ibid., 176, 183.

24. Ibid., 176.

IS TECHNOLOGY THE FRUIT OF CHRISTIANITY?

The environmental crisis and Christianity are often linked because *technology*, as fruit of Christianity, is held especially responsible for the current deplorable situation. That is also the most important point in White's argument, as we have seen. But is it legitimate to make this link?

Strikingly, the link between Christianity and technology was considered a *positive* one before people became aware of the environmental crisis. For example, A. Th. van Leeuwen and H. Berkhof regarded science and technology as fruits of the liberation from deifying nature and from magic and myth, so that people could open themselves up to (research and use) nature. Because Christians do not view nature as divine, but "only" as a creation, it was understandable that technological intervention in nature could take great strides with Christianity.[25]

In fact, this idea is older than might appear from the books of the two authors mentioned above. Although Berkhof did not realize it when he wrote *Man in Transit*, he was in fact echoing what Abraham Kuyper had already argued in detail at the turn of the twentieth century.[26] It is useful to reproduce Kuyper's argument here.[27]

According to Kuyper, the enormous power that people in the Christianized world acquired over nature derives from the natural sciences, which did not flower in China, Japan, India, and Turkey like they did in Christian Europe and Christian America. Kuyper even dared to posit with certainty that we can see a blessing of the Christian religion in the power of these natural sciences.

---

25. Van Leeuwen, *Christianity in World History*; Berkhof, *Man in Transit*. A clear example of the high regard for modern technology can be found in Cox, *Secular City*.

26. See my *Algemene Genade*, 59; Kuyper, *Common Grace* (1.1: Noah–Adam; 1.2: Temptation –Babel; 1.3: Abraham–Parousia); *De Gemeene Gratie* II-III; and *Pro Rege* I-III. Kuyper wrote the articles that are collected in *Common Grace* and *De Gemeene Gratie* in his newspaper, *De Heraut*, between 1895 and 1901.

27. I discuss Kuyper in similar terms in my *Algemene Genade*, 51–59.

Environmental Stewardship

Why? In Paradise, God told man that he would have dominion over nature. We lost that dominion in the fall. But through God's common grace much of that dominion over nature was given back to us. We can see that in China, India, and Islamic countries, whose power and dominion over nature rises far above that of the African nations. This common grace is tripled in power through the church of Christ and particular grace. Only through Christianity are people truly liberated, so that they can again exercise full dominion over nature.[28]

Kuyper connects the *Christian* character of dominion over nature directly with Christ himself. In him people reach their fulfillment in the richness and height of his power. In Christ Psalm 8 is fulfilled—the psalm that speaks of man's dominion over nature. Kuyper sees the evidence for this in the *miracles* that Jesus performed during his time on earth. He performed these not only as God but also, and more particularly, as restored man. Jesus' power to drive out demons, heal people, calm storms, etc. is the evidence of what restored people are able to do.

What Jesus did we can no longer do, of course. The *immediate* power to perform miracles, which enjoyed a late but limited flowering in his apostles, ended with Jesus. But, says Kuyper, people's *mediate* power to perform miracles does not end, for they triumph over nature "through the higher development of spiritual factors in our race." Indeed, we have the prospect of greater power than Jesus himself exercised.

Kuyper relies on John 14:12 to support this assertion: "Truly, truly, I say to you, whoever believes in me will also do the works that I do; and *greater works than these* will he do." Thus, in their research and meditation people would be able to penetrate the essence of nature and thereby discover the hidden powers it contains, in order to make those powers serve humanity's dominion over nature.

It is impossible to understand this apart from Christ, according to Kuyper: "If you eliminate Jesus' appearance and the entry of Christianity into Europe, we should be, like the Chinese in the

28. Kuyper, *De Gemeene Gratie* II, 274–75.

East and the Indians in the South of Asia, just as powerless as the ancients were over against nature." Christ's gospel has produced a totally different and a much higher development of people's spirit. The mediate power that has been given to us operates in all countries and peoples, century after century, "and blesses thousands at the same time in all distress and illness."[29]

TECHNOLOGY, A GENERAL PHENOMENON

Clearly Kuyper was impressed by the scientific and technological achievements of his day when he wrote these lofty words. Yet, as a result of the environmental crisis, we have begun to see the downside of people's mediate actions more clearly, so we can no longer speak of the blessing of science and technology "in all distress and illness." Accordingly, we need to take a critical look at the real or perceived link between Christianity and technological progress.

Is it true that technology and Christian faith should be conjoined so exclusively? Even in pre-Christian cultures people achieved substantial technological success. They learned how to control the water supply at the mouths of the Nile and the Indus, and in the territory between the Tigris and the Euphrates Rivers, by digging canals and constructing dikes, for instance. Apparently religion did not prevent people from creating arable land by means of impressive irrigation works.

The invention of the plow, the wheel, mine development, metal work, agriculture, the science of surveying, and the making of glass and enamel ceramic products, canoes and small inland navigation vessels, the winch, the pulley, and mill stones occurred thousands of years before the beginning of Christianity. For the Jerusalem temple, Solomon used builders from pagan Tyre. Hiram, the metal worker, also came from there (1 Kgs 5:7, 13–18). The Greeks and Romans further developed what had been invented in the neighboring Eastern countries. Many technologies (including

---

29. Kuyper, *Pro Rege* I, 183–84, 186–87.

central heating, which the Romans used already) were lost for a long time after the Roman Empire fell.

Whether technological development occurs does not depend ineluctably upon a certain religion. Jacques Ellul, for example, has pointed out that it is possible to discern a kind of technology in pagan *magic*. Practitioners of magic try to control the powers of nature by performing specific rituals. As far as the Christian faith is concerned, Ellul defends the proposition that Christianity actually restrained technological development for centuries. Were Christians not supposed to focus on the spiritual world rather than the physical one? Thus, the view that nature is not divine does not necessarily mean that technological development is the inevitable result. Indeed, Ellul posits that the Christian viewpoint in the ancient church and the Middle Ages actually served as a brake upon rather than as a spur behind technology.[30]

All kinds of knowledge about and skill in technology were developed and preserved in an Arabic-Muslim, not a Christian, society. In addition, Muslims made use of inventions from China, such as the derivation of sap from sugar cane and the production of paper.

It is possible, of course, to posit a link between the invention of the eight-oxen plow in the ninth century and the Christian view of nature, as White did. But it is equally possible to argue that the waning of serfdom within the economic system made people technologically creative so that they could work more efficiently in a cooperative way. Watermills and windmills also came to Christian Europe, but only after the Arabs had already determined their principle and application.[31]

---

30. Ellul, *Technological Society*. For a helpful overview of Ellul's vision concerning the development of technology, see Schuurman, *Technology*, 125-58.

31. See Aninga, "Techniek," in *Grote Winkler Prins Encyclopedie*, 8th ed., vol. 21, 475; Houtman, *Wereld*, 74. Ellul denies the significance of the abolition of slavery for technology. He points to Egypt, which had many slaves and a highly developed technology, in contrast to Israel, for example, which did not have widespread slavery but also had little technological development (*Technological Society*, 35-36).

Environment

As soon as we enter modern times, which were influenced not only by the Reformation, but also by the Renaissance and the Enlightenment, we can justifiably ask ourselves which "belief" is really the basis of modern technological advancement.

René Descartes (1596–1650) developed a philosophy based upon a duality between the mind as a thinking, non-extended substance and the body as an extended, non-thinking substance. The human spirit, as thinking substance, rises far above nature. Nature, as extended substance, can be measured and weighed. On the basis of his theory, Descartes asserted that we can employ the power and functions of fire, water, air, the stars, the heavens, and all other bodies that surround us, for all uses to which they are suited, "in the same manner as the occupations of our artisans." Indeed, such power makes us "lords and masters of nature."[32]

When you read this and realize that Descartes regarded even animals as *machines*,[33] you can reasonably ask a number of questions: When technology justifies itself by such philosophical arguments, is technology still the fruit of the Christian faith, or does it derive from apostate thought? Can you draw a line from Christianity to modern technology, or is technology based in the rationalism of the Enlightenment?

---

32. Descartes, *Discours*, 61–62. Gilson, the text's editor, says in his commentary on this passage (among other things) that Descartes would have enjoyed our era, which is marked by engineering triumphs, and nothing would make him more passionate than the ever increasing number of *machines*, which people use today to look after and transport themselves, to make their products, and to communicate with each other! (ibid., 444).

33. See ibid., 420ff. Of course, this does not necessarily imply that Cartesians torture animals. For Descartes and associates, an animal is not *merely* a machine that you can therefore treat as you wish, but it lives conditioned *as* a machine. According to Descartes, the same applies to a person's bodily functions. Francis A. Schaeffer uses the example of the group of people the Dutch call "black-stocking Christians." Many of them supposedly beat and torment their livestock because animals do not have a soul or a heavenly destination and therefore do not have to be treated kindly. See his *Pollution and the Death of Man*, 41. Such a conclusion is as unfair as the one about Cartesians torturing animals. On that basis one can "prove" that ultra-Reformed people beat their wives often, because in their view Christ has not elected most of them.

Environmental Stewardship

Kuyper sensed that problem too, for he asked why general human development has normally been borne by unbelievers rather than by believers. In his answer he identifies the struggle between the spirits hidden from human perception. After the Holy Spirit was poured out, the demonic forces were "driven back by a holy spiritual atmosphere." The influence of that extended beyond the church, and when Christians withdrew from their task to investigate nature, especially unbelievers took over the inquiry that the Christian community had neglected. When demonic powers are at work, however, King Jesus keeps them in check and fights them.[34]

But this answer is not very convincing, especially when we look at the text Kuyper cited. It speaks clearly of the actions of people who *have faith in Christ* (John 14:12)! Kuyper's argument is exceedingly speculative. It is better to admit that the works ("things") mentioned in that verse do not have anything to do with scientific or technological achievements.

CONCLUSION

We do well not to assign a key position to Christians in the development of science and technology. Christians and non-Christians alike have their place in that development, and in this respect non-Christians may be more impressive than Christians. It has been that way from the beginning: the first people who excelled in manufacturing and playing musical instruments and working metal were children of Cain, not of Seth (Gen 4:21–22).

We must not reduce the development of technology to one factor, no matter how fascinating such "simple" theories may be. We have to take into account a multiplicity of complex factors instead. Greek rationalism, the technology of the near and far East, the work of engineers, the travels and commerce of the Romans, the humanism of the Renaissance, the rationalism of the Enlightenment, the reliance on humanity itself, economic and

---

34. Kuyper, *Pro Rege* I, 188–94.

societal factors—all these must be considered *together* with the influence that the Bible's vision has exercised over nature in the total process.[35]

Accordingly, I can be quite brief about the claim that Calvinism in particular is responsible for environmental pollution. Elsewhere in this series, entitled "Ethical Reflections," I have discussed the Max Weber thesis.[36] Forms of capitalism did not arise in every Calvinistic region, contrary to the impression that Max Weber gives. Nor is it true that all countries influenced by Calvinism experienced accelerated technological development. The problem is too complex to blame everything simply on Christianity or Calvinism.

Technology, including highly refined technology, does not flourish only in a Christian or, more specifically, a Calvinistic culture. Japan serves as an apt example, for we certainly can no longer say about Japanese technology what Kuyper did in his day: "They copy us."[37]

The converse is also true. Technology has advanced in every Christian territory. Besides, Christianity is a multiform phenomenon, and thus it is difficult to draw a general conclusion about the effect of the Christian faith on technology.[38]

35. See Derr, *Ecologie*, 38–39.

36. See my *Vrede in de maatschappij*, 46–50, 140.

37. Kuyper, *Pro Rege* III, 348. Elsewhere he wrote, "When one considers recent developments in Japan, one is struck by the strange phenomenon of what is still a real pagan society, which is overlaid by the shine of European veneer. . . . There is no organic cohesion between what they themselves have developed and what they adopted from Christian Europe" (*De Gemeene Gratie* II, 179). Meanwhile, it has become quite clear that technology has flourished in that pagan soil just as much as in Christian soil (or whatever passes for it). The influence of Buddhism is cited to explain why individualism is much less prominent in Japan than in Western industrialized countries. The Japanese pay greater attention to harmonious relations in business, which functions like a family. See Huppes, *Ambachtelijk elan*, 111.

38. Eastern Orthodoxy developed a mystical society in Russia in which people were indifferent toward the material world and did not embrace science or technology as people did in the Latin West. But was this the result of doctrine? Or can one agree with Ellul that it was the result of the national temperament, rather than of Christian doctrine as such? See Ellul, *Technological*

Environmental Stewardship

Therefore, in my opinion, when you look for those responsible for the current environmental crisis, you simply find *humanity*. It is impossible to assert that Christians, or even more specifically, Calvinists, are especially responsible. Christians are not innocent; that is clear. It is also clear that environmental pollution in the rich Western countries, which owe so much to Christianity, has reached drastic proportions. Wealth and waste at the cost of the environment are going hand in hand. Christians must certainly search their own hearts. But that is different than ascribing the environmental crisis to Christian and Calvinistic principles, and certainly to what the Bible says about the environment.

That is what we will discuss in the next chapter.

Society, 32–38; and Derr, *Ecologie*, 37.

2

The Bible and the Environment

WORLD AND ANTI-WORLD

EVEN IF WE ACCEPT that Christians and, more specifically, Calvinists have caused the environmental crisis, this does not mean that the *Bible* set them on this track. Therefore, we now have to consider whether Jacobi and others are correct when they discover environmentally hostile pronouncements or tendencies in the Bible itself.

I should like to point out first that it is easy to give a distorted picture of what the Bible describes when we read *today's* circumstances into biblical texts that are thousands of years old. It has been said, correctly, that pleas for environmental protection are to be expected from people who come into contact with only sparrows and sleeping dogs in their daily lives, and who know wild animals only from pictures or a visit to a zoo.[1] If you read the Bible from that perspective, you will undoubtedly come across all kinds of statements that seem to permit destruction of the environment.

---

1. Houtman, *Wereld*, 31.

Environmental Stewardship

On the other hand, merely citing a few Bible texts, such as Psalm 8, Psalm 104, Job 38–41, and Matthew 6:26–28, which pay particular attention to God's good creation with its animals and the beautiful "lilies of the field," will not suffice. For it is true that, both in the criticism of the Bible and in its defense, people often lose sight of the fact that the Israelites, as well as the pagans in neighboring countries, were engaged in a constant struggle with the natural environment. The ground was cursed because of humanity's fall into sin and it brought forth thorns and thistles (Gen 3:17–19).

In biblical times "the environment" did not yet constitute a distinct topic we might expect to be discussed in the Bible. The protection of nature in the sense of preserving flora and fauna is not an issue when you have to struggle *against* thorns and thistles and wild beasts.[2]

Thorns and thistles may *now* seem attractive in Palestine; *then,* people who had to toil to earn their daily bread did not appreciate them. The fact that some twenty words in the Old Testament are translated as "thorn," "thistle," "nettle," and "weeds" already indicates that the world we call "nature" is also an "*anti*-world," to use Cornelis Houtman's term.[3] This anti-world is an enemy of human beings. Thorns and thistles are vile plants that even serve as metaphors for enemies and the ungodly (Num 33:55; Josh 23:13; Ezek 28:24).

Of course, the Bible also favorably regards certain trees, plants, and crops, which are thus distinguished from the thorn bush. Some of these include the olive tree, the fig tree, the grape vine (Judg 9:8–15), the cedar (2 Kgs 14:9), myrtle (Isa 55:13), grains (Job 31:40), and the lily (Song 2:2). The Israelites had respect for large trees with longevity and strength, such as the cedar, cypress, and oak (Isa 65:22; Ezek 17:23; Amos 2:9), and they were conscious of the splendor of the cedar, the plane tree, and the pine, which beautified the temple (Isa 60:13).

---

2. I did not give due attention to this aspect in my earlier publication about the topic, "Milieu en Heilige Schrift," in *Het gelaat van de aarde*, 27ff.

3. Houtman, *Wereld*, 31.

The Bible and the Environment

But nowhere do we find statements suggesting that the protection of trees and plants is an end in itself. The cedars of Lebanon were impressive, but they were cut down for the construction of the temple and houses (1 Kgs 5:6–18; Song 1:17). The Bible does not record any objection to the clear cutting of Lebanon, as such. It protests only when the lumber is destined for *improper use*, such as godless self-exaltation, the manufacture of war materiel, or the construction of fortresses in which tyrants think themselves safe (Hab 2:17; 2 Kgs 19:23).

The Bible depicts trees and plants solely as a *means of sustaining life*. The Israelite's attitude toward fruit trees is typical of this depiction. During the siege of a city, Israel was entitled to cut down all trees except fruit trees (Deut 20:19–20) because, when the siege ended, the inhabitants of the city would need the fruit for their sustenance.

This data demonstrates that the ideal world for the Israelites was not one in which people defend the rights of unspoiled nature or where people are supposed to strive for ecological balance. Instead, they longed for a world in which the wilderness is recreated as a forest with trees they can utilize (e.g., Isa 41:18–20; 55:13). Only fruit trees are mentioned in Paradise (Gen 1:11–12, 29; 2:9, 16; 3:22; cf. Rev 22:2). The ideal was a world that produces abundant fruit for the people (Jer 31:12; Ezek 47:12; Joel 3:18; Amos 9:13–15). The earth is then to be turned into *arable land*, so that grain will grow abundantly even on the tops of the mountains.[4]

We can draw similar conclusions about the *fauna*. The messianic kingdom is depicted with images of wolf and sheep, leopard and young goat, cow and bear, and weaned child and adder living together in peace (Isa 11:6–9). Humanity's struggle with the anti-world, which is also manifested in the animal kingdom, will end (Hos 2:17). Like thorns and thistles, wild animals endanger people, who then have to respond to the danger with force, as Samson and David did when killing a lion and a bear (Judg 14:5–6; 1 Sam 17:34–37). Also, just like the thorns and thistles, wild animals can

---

4. Ibid., 26–27.

Environmental Stewardship

be symbols for the ungodly and enemies (e.g., Ps 22:13, 22; Isa 56:9–12; Jer 2:15; 4:7; 12:7–8; Ezek 39:17–20).

It is different, of course, with the animals that people can use for their own benefit, such as the sheep, goat, cow, and donkey. They can enlist these into their service as domesticated animals.

Nowhere does the Bible draw attention to the protection of animals in the way we do now for the protection of nature. It is true that the Israelites were not permitted to muzzle an ox when it was treading out the grain, so that it could eat of the grain while it did its work (Deut 25:4). Further, the Bible tells us that the righteous have regard for the life of their animals (Prov 12:10). But these references concern the protection of domesticated animals, not the protection of animals *in general*.

Perhaps the directive to let the mother fly away when you remove the young or the eggs from the nest (Deut. 22:6-7) suggests a motivation of "humane motives."[5] But it is also possible that a motive of utility underlies this command. If you killed the mother, she would no longer produce more eggs or young, which people use for their food.[6]

The rest prescribed for the animals on the Sabbath (Exod 20:10, Deut 5:14) and for the land during the Sabbath and Jubilee years (Exod 23:10–11; Lev 25:1–12) is not by itself proof of any special attention for the environment. *People* had to rest from their labors on the Sabbath day and in the Sabbath years, and to make *their* rest complete they had to also let their animals rest. Unsurprisingly, disobeying the command to rest on the Sabbath and to observe the stipulations for the Sabbath and Jubilee years amounted to a transgression of religious law that would result in disasters for the *nation*. We do not read anything about the injury to animals and the earth caused by the people's transgression (e.g., Neh 13:18; Ezek 20:13; 2 Chr 36:21).

---

5. See, e.g., Von Rad, *Deuteronomy*, 141.
6. See Phillips, *Deuteronomy*, 146.

GENESIS 1:26-30

With this information we have not yet provided a complete picture. Similar biblical examples exist, and we will have to consider other matters as well. But before I do so, it is useful to pay attention to Genesis 1:26-30. From the biblical information mentioned above, *people* clearly occupy center stage, even when plants and animals are mentioned. There is a huge difference between what the Bible teaches, and what pagan religions proclaim about the divine nature in all living things, so that trees and animals can be declared holy and consequently inviolable. The Israelites may not have consumed many kinds of animals, but this prohibition was based on the uncleanness and not on the holiness of those animals. They were permitted to eat clean animals and use them for the sacrificial service of Yahweh (Lev 1-7; 11:46-47).

At creation, God gave human beings the important position they occupy among other creatures. Genesis 1:26-30 is very clear about this. God created human beings in his image, something not said about any other creature. As God's image, people are God's *representatives* on earth.[7] God is king; human beings are his vice-regents. That position gives people the right to "have dominion over the fish of the sea and over the birds of the heavens and over the livestock and over all the earth and over every creeping thing that creeps on the earth" (v. 26). Thus there is no equality: people have dominion over the animals. But does this important fact necessarily imply that the environment, transferred into Jewish-Christian hands, must suffer? That depends on the character of this *dominion*. I do not think it is right to read a *harsh* tone into the "dominion" outlined in Genesis 1.[8] Dominion is harsh here only if you already read the fall into sin, which is first described in Genesis 3, into Genesis 1. If you do that, then human beings in Paradise already had to struggle and act harshly to conquer the wild animals and maintain their place in unruly nature. When

---

7. I have provided a more extensive explanation of this designation in *The Ten Commandments*, 50-54.

8. See Houtman, *Wereld*, 41.

they received dominion, human beings also received the ability to *regulate the conflict* that could easily arise between themselves and the animals.[9]

But Genesis 1:28–30 certainly does not point to a situation of (possible) tension and struggle in the unfallen world. The fear and terror of animals toward human beings is not even mentioned until Genesis 9, in connection with the stipulations of the so-called Noahic covenant. At that time, too, people were permitted to use animals as food, while in Paradise they were allowed only the fruits of plants and trees, and the animals were also restricted to eating plants (Gen 1:29–30; 9:2–3).

The "dominion" of Genesis 1 was, therefore, a harmonious reign.[10] God had declared Creation to be "very good" (Gen 1:31). That is why the animals paraded in peace before Adam to receive their names (Gen 2:19–20). In their dominion, human beings showed themselves to be the image of God. And why should the exercise of that dominion not be just as peaceful as that of God over people?

In addition to "dominion," Genesis 1 uses the verb "subdue." What needs to be subdued? The earth, because people need to obtain food by tilling the ground (Gen 1:28).[11] Their assignment thereby encompasses culture in the most original sense of the term. The first meaning of the Latin verb *colere* is "tillage of the land."

9. Thus Liedke, *Im Bauch*, 130ff., who perceives a "Konfliktreglung" (conflict regulation) in the *dominium terrae* (dominion over the earth). The Hebrew word used for "dominion" (*rdh*) often has a harsh tone because of the context in which it is used. But it can also have another sense. See, for example, Leviticus 25:43 and 46, which show that slave owners may exercise dominion, but not ruthlessly!

10. Also in the apocryphal literature, the dominion of human beings over animals in Paradise is regarded as dominion "in solidarischer Gemeinschaft und nicht in tödlicher Feindschaft" (in solidary community and not deadly enmity), as Grässer demonstrates with several citations in "Kai èn meta toon thèrioon," 144ff.

11. Steck, *Welt*, 79ff., 180, on the basis of the information in Genesis 1 and 2, rejects every negative undertone and ambivalent quality in the subjugation and the exercise of dominion by human beings (in Paradise).

The Bible and the Environment

RESPECT FOR ANIMALS AND PLANTS

We have to examine Genesis 1 and 2 carefully and may not, therefore, interpret the words "dominion" and "subdue," used in the context of Paradise, as though they already warn of a threatening *conflict* between human beings and their environment. When we do *not* do that and instead interpret the original exercise of dominion as peaceful and harmonious, we understand more clearly that the Bible is not solely concerned with human beings and their struggles. Let me provide a few biblical examples.

Human beings have been created in God's image, but they themselves are not God. Therefore they cannot do whatever they want, as becomes clear already in Genesis 3. They can always be called to account by God.

While human beings were created in God's image, God created plants, trees, and animals *according to their kinds* (Gen 1:11–12, 21, 24–25). That means at the very least that they received something unique to their natures. Human beings are permitted to have dominion over everything, but in doing so they must respect the unique nature of all other living creatures.

Human beings received the task to work and *keep* the Garden of Eden (Gen 2:15). "Working" the garden operated in tandem with "keeping" it, that is, guarding and preserving it. In the first place, guarding is probably directed against the invasion of animals, which could become instruments of a seductive, devilish power (Gen 3:1–15). But the directive to preserve the garden also always means to keep it intact as a creation of God that may not be destroyed by anyone or anything. Indeed, the earth, people, animals, and fruit must be preserved in the state in which God created them.

When the flood threatened all the animals on the earth, God permitted a selection of them to survive in the ark (Gen 6:19–20). After the flood, God entered into a covenant with the people *and* with "every living creature that is with you [Noah and his family], the birds, the livestock, and every beast of the earth with you, as many as came out of the ark" (Gen 9:9–10). The Noahic covenant is central to any discussion of the environment, and rightly so. God, who promised that he will never again send a flood to destroy the earth, is sympathetic not only toward people, but also toward animals.[12] That is why we must pay attention to Yahweh's separate attention to animals in the Fourth Commandment, in which we are commanded to let the *livestock* rest as well. And when God permitted people to use animals for their food after the flood, he imposed a limit: they may not eat an animal's blood, because the blood is its life, the essence of the animal's being (Gen 9:4).

Human beings occupy a high position in the world; they have dominion over everything (Ps 8). But God can also show us how small we are by drawing our attention to animals like the lion, wild donkey, hippopotamus, and crocodile (Job 38–39), so that Job eventually says, "Behold, I am of small account, what shall I answer you? I lay my hand on my mouth" (40:4). God's care and attention encompass the entire world. He is not intensely involved only with human beings, but also with trees and animals (Ps 104). Solomon in his wisdom also concerned himself with trees, from the cedars of Lebanon to the hyssop, and with livestock, birds, reptiles, and fish (1 Kgs 4:33). God spared Nineveh not only because of the 120,000 people who lived there, but also because the city contained much cattle (Jonah 4:11).

While the rest prescribed for the land during the Sabbath and Jubilee years was not based on environmental considerations first of all (see above), we should nevertheless take note that anything that ripened on the land during those years was designated for the poor and the animals (Exod 23:11; Lev 25:7).

People can derive useful lessons from the lives of animals and plants such as the sluggard and the ant (Prov 6:6–8), and those who are anxious can learn from the birds and the lilies (Matt 6:25–30). Ants, rock badgers,[13] horses, ostriches, donkeys, dogs, migratory birds, mice, and swallows all fulfill a role in the preaching of the Bible.[14]

---

12. See Blok, "Met reikhalzend," 276ff.
13. *Translator's note*: Also referred to as coneys and hyrax.
14. Smelik, *De ethiek*, 154.

The images that describe the messianic kingdom remind us of a restored Paradise, in which animals and people live together without fear and conflict—wolf and sheep, leopard and young goat, cow and bear. A child can lead them and a baby can place its hand in the adder's den (Isa 11:6–9). Something of that kingdom could be seen during Jesus' life, particularly when he was tempted for forty days in the wilderness by Satan. It was not Satan, but Jesus, who "was with the wild animals" during this time, who was lord and master of that environment (Mark 1:13b). He restores peace in creation![15]

The conflict between people and their *anti*-world is quite apparent in the Bible, as we have seen. But it is also clear from the Scriptures that people's attitudes toward other creatures must be one of respect, for they have their own place in God's creative and caring work. The conflict began after the fall into sin, but that is not how it was in the beginning. Dominion is not fulfilled in exploiting and destroying flora and fauna, but in respecting and caring for them.

ANTHROPOCENTRIC OR COSMOCENTRIC—AN INCORRECT DILEMMA

With the above-mentioned information in mind, it is not difficult to find a solution in the Bible to the following false dilemma: some people say that our environment has been destroyed by those who hold an *anthropocentric* vision. In this philosophy of life people occupy a central place and the rest of the world is subordinated to them. Instead, others argue, this philosophy of life must be replaced by a *cosmocentric* vision, in which all of nature receives due attention and consideration.

Confronted by this choice, we reject both possibilities. It is true that people must live together with animals, plants, wind, water, earth, and sky. But the cosmocentric vision goes a step further than simply recognizing this interrelationship. It regards human

---

15. See Grässer, "Kai èn meta toon thèrioon," 152ff.

life to be on par with the other forms of life. The adherents of this vision appeal to "respect for life," as Albert Schweitzer (1875–1965) wrote about it, and thus reject all ethics that give special consideration to human life. They call ethics of humanity discriminatory.[16] A more pantheistic version of cosmocentrism deems all of nature divine, and people as just one expression of that divinity.

On the basis of what the Bible says about people as the image of God, it is impossible to agree with the cosmocentric vision. People have a unique place in the world, which is clear when we pay attention to their relationship to God and their position above other creatures. Moreover, people themselves know that they are different, special, and unique, even if they hold to the cosmocentric vision. Their cultural artifacts, for instance, witness to their otherness. It is not arrogant to insist that people are unique and occupy a special place among all creatures; rather it is stating the obvious.[17] Ecology also is *human* ecology.

At the same time, rejecting all forms of cosmocentrism does not mean that we therefore must adopt an anthropocentric vision. For the dilemma is a false one. There is a third possibility that does justice to *all* creatures. We can call this possibility the *theocentric* vision. People do have the lead role, but they are in turn subordinate to God, who made all creatures "according to their kinds." This indicates that there is a differentiation in creation that people have to respect.

The phrase "according to their kinds," which describes the variety of both plants and animals (Gen 1:11–12, 21, 24–25), indicates that plants and animals lead their own lives. Their life is comprised of their God-given ability to propagate themselves and multiply in their diverse kinds. Creation is an ordered whole, not an undifferentiated mass.[18]

People are not masters over "things," which they may handle and transform as they please for their economic enrichment;[19]

16. Bouma, *Slagveld*, 14.
17. Thus, correctly, Derr, *Ecologie*, 26–27.
18. Keil, *Pentateuch*, on Gen 1:11; Westermann, *Genesis 1–11*, ad loc.
19. Wiskerke, "Meeleven," 108: "The 'point' of modern anthropocentrism

rather, they receive stewardship over a pluriform creation, in which plants, animals, and people have their own worth, determined by God. Thus it is clear that people must treat other creatures in accordance with the aptitude and function that God has given to each.

The world is God's creation and therefore cannot be regarded as the possession of the *human beings* who occupy it. People are not lords *over* creation, but only lords *in* creation.[20] Creation belongs to God and he only loans it to human beings to govern in accordance with divine justice, not in accordance with delusions of grandeur.[21] Creation does not revolve around human beings and their power. They do occupy a central position in creation, but it has been rightly said that we may not call them "the crown of creation." God created the world for his own glory and the crown of creation is therefore God's *Sabbath* (Gen 2:2–3). Human beings, together with all earthly and heavenly creatures, bring praise to God and share in the enjoyment of God's Sabbath. The heavens declare the glory of God, even without human beings![22]

When we experience the world as the Creator's gift to human beings, and when we behave ourselves in reverential amazement, humble modesty, and God-praising gratitude,[23] no creature will be shortchanged.

---

finds its true definition not so much in the conviction that nature serves as a commodity for the human being and that the latter must be regarded as the most important being in all of nature, but only in a reduction of the human–nature relationship that treats a person as *homo faber* [worker], for whom nature principally appears in the extremely one-sided network of production, production room, and material for work."

20. Friedrich, *Ökologie und Bibel*, 15, who cites Barth, *Church Dogmatics*, vol. III/1, 206–7.

21. Moltmann, *God in Creation*, 20–40.

22. Psalm 19:1. See also ibid. A good celebration of the Sunday (which Moltmann does not regard as an extension of the Sabbath, however [ibid., 276–96]) is also a blessing for the environment.

23. Auer, *Umweltethik*, 203. For this reason, it seems to me better to avoid using the term "Anthropozentrik" (anthropocentric), which Auer continues to use.

Environmental Stewardship

STEWARDSHIP

When we take into account the above-mentioned texts from the Bible, an accurate description of the attitude human beings must have within creation begins to take shape.

All kinds of characterizations are in vogue to describe the attitude of human beings toward other creatures. Many people follow in the steps of Francis of Assisi (1181/1182–1226), who in his *Canticle of the Sun* addressed sun, moon, stars, wind, water, fire, and earth by turns as "brother" and "sister," and often preached to other creatures.[24] Others also use terms such as "friend" or "hostess" to express the personal relationship between people and nature. Clearly, those who use such terms find it necessary to place people and nature in a triangular relationship with God, in which nature functions as humanity's full *partner*.[25]

But the Scripture passages that we have considered do not permit us to speak of human beings as equal partners with animals and other creatures. Nature has not been given its own *independent* place alongside God and human beings. The Bible does not accord nature an independent status, not even in the so-called nature psalms, like Psalm 104. It has been rightly said that Psalm 104 is not a paean to ecology and that Romans 8:18–23 is not about an ecological redemption of creation from its groaning.[26]

Non-human nature has its place, of course, but that place is always dependent upon the relationship between God and human

---

24. For the Latin text and translation, see Maximilianus, *St. Francis' Zonnelied*. Maximilianus' introduction demonstrates that Francis' love of nature had a religious basis. The creatures are shadows and paintings of religious truths, just like the lamb for the Lamb of God, wood for the cross, the rock for Christ, the cornerstone, etc. Francis also sometimes referred to the animal as a moral example, or as a term of abuse. His attitude toward nature has little or nothing to do with the modern concern for the destruction of the environment. Besides, Francis was very much aware of the special position of human beings. For further information, see Verheij, ed., *Franciscus van Assisi*; Houtman, *Wereld*, 64ff.; and Verheij, "Franciscus," 10ff.

25. See, e.g., the report of Berkhof, "God in Nature and History." For other examples, see Houtman, *Wereld*, 42ff.

26. Derr, *Ecologie*, 44.

beings, which is decisive for everything and everyone. That is why I maintain, on the basis of the Bible, that we must continue to recognize only two basic relationships, not three: human beings have a relationship with God and with their fellow human beings.[27] We should not add "nature" to these relationships, either as "brother," "sister," or "friend."

A better characterization exists that does justice both to the special position of human beings and the significance of the other creatures. I am thinking of the term *steward*. This designation clearly portrays the theocentric vision. God is the owner; human beings are but stewards tasked with caring for the Lord's possessions.

The care and management of these possessions must be both responsible and efficient. For that purpose, the term "steward" is appropriate in this context. This is especially apparent when we prefer it to the term "tenant." A tenant is someone who, after paying the specified rent, has control over the land and its yield. He is entitled to whatever he can generate from the land beyond the rent. He can exploit the land after he has paid the rent. But stewards must be careful managers, so that the goods entrusted to them remain in, or reach, a state desired by the *owner*.[28]

Yet every term has its limitations. The term "steward" connotes one thing to one person and something else to another. Some may be reminded of the economic entrepreneur, for whom everything must be profitable, leaving little room to enjoy what has been given us in nature. It has also been said that the preference for the term "steward," especially among Calvinists, betrays a one-sided emphasis on nature as servant and enemy and an underestimation of nature as a friend.[29]

27. See my *Responsible Conduct*, 36–37.

28. See also my volume *The Ten Commandments*, 295–306.

29. Thus Jager, *Schrale troost*, 66–67. See also Manenschijn, *Geplunderde aarde*, 170ff., who cites an article by W. Coleman. Coleman argues that some theologians in England in the seventeenth and eighteenth centuries interpreted the stewardship tradition as economic-individualistic. The human entrepreneurial drive was justified theologically as a stewardship that was despotic! However, others have disputed Coleman's conclusions. Manenschijn refers

Environmental Stewardship

I think that this criticism is beside the point. "Steward" is a useful watchword when we compare it to other concepts, such as human beings as "sisters" or "friends" of nature. Such designations negate the unique position of human beings between God and other creatures, whereas "steward" makes that position clear. On the one hand, the term indicates that human beings exercise supervision over their fellow creatures, while on the other hand, a steward is merely a steward, who has to manage the owner's property in accordance with the latter's wishes. All creatures, which God has made "according to their kinds," must be managed by the steward in accordance with their natures.

The term "steward" includes the concept of *profitability*. That can have negative connotations for those who espy "capitalism" everywhere. But if stewards engage profitably with nature, their actions need not necessarily be exploitative. Indeed, we are permitted to make nature profitable for our own use. But the phrase "to work it" in Genesis 2:15 is irrevocably tied to the phrase "and keep it." Exploitation is never profitable. If we strive for beautiful rivers, healthy forests, and a clean atmosphere today, we are being profitable. A good environment is good for our well-being too.

Profitable stewardship is also directly relevant to our faith. In article 2 of the Belgic Confession we read that the Bible is a most elegant book, in which all creatures, great and small, allow us to perceive clearly God's "invisible attributes, namely, his eternal power and divine nature" (Rom 1:20). Good environmental management thereby provides opportunity to praise God the Creator. We should take this into account when we consider how we can work profitably as stewards under God. When industries pollute the earth for their own advancement, the profitability that we gain as stewards of creation is nullified. A polluting steward obscures our view of the great Owner who wants us to enjoy his creation.

---

to R. Attfield, who calls Coleman's conclusions one-sided and unreasonable. While the despotic tradition certainly existed then, there was also a vision of human beings as stewards with respect for nature.

AMPLIFICATIONS

A watchword like "steward" can elucidate and amplify other concepts as well. Bockmühl, for example, places stewardship, contentment, and care alongside each other. We are stewards who have received managerial ownership from God, but not an ownership in the sense of the *dominium*, as found in Roman law, which allows you to do with your property whatever you wish. Moreover, we must show *contentment*. Greed is the root of all evil, so we must be satisfied if we have sustenance and shelter (1 Tim 6:8–10). Also, we must exhibit *care*, according to Bockmühl. Our management must be directed toward our neighbor. When they preached about the maintenance of property, for instance, the church fathers tended to focus less often on the parable of the talents than on the words of Christ in Matthew 24:45–46: "Who then is the faithful and wise servant whom his master set over his household to give them their food at the property time?" People are not only highly placed stewards, but also slaves or servants. The steward does not profit from the chattels under his management; rather, he distributes property to those in need.[30]

In my opinion, these amplifications illustrate how we must conduct ourselves if we want to be *good* stewards. We must not abandon the term "steward."

CULTURAL MANDATE

The term "stewardship" is often coupled in Christian circles with the phrase "cultural mandate," which is linked to the dominion over and subduing of creation in Genesis 1:28. While considering the cultural mandate, we may come across the expression that "everything contained in the world must be extracted from it." We read that, for example, in the well-known book of K. Schilder, *Christ and Culture*. But does not the environmental crisis make it

---

30. Bockmühl, *Umweltschutz*, 25ff.

clear that we must instead stop extracting from the world everything that it contains?[31]

Let us first determine what Schilder meant when he used the idea of cultural mandate. He was not the first who linked this subject to Genesis 1:26–27, for Kuyper had already done that. Both men argued that Genesis 1 formulates a cultural task for human beings. The fact that Kuyper spoke of a "cultural task" and Schilder of a "cultural mandate" makes no difference. But I now restrict my arguments to Schilder, who in *Christ and Culture*, which is still widely read, writes extensively about the cultural mandate. What does he say about it?

Schilder speaks about the culturally interested Creator, who placed human beings in a world that was not yet finished. Culture is an idea already present on the first page of the Bible: cultivate the garden, fill the earth, be fruitful. Paradise was but a beginning, a world of potentialities that could be developed into a completed world. That development would require a span of many centuries. We should not imagine Paradise as a magical enclosed place surrounded by gardens; rather, it was the beginning of the *adama*, that is, the inhabited world. The garden lay open. Creation must advance—via evolutionary development—toward a completed culture.

All these elements come together in Schilder's lengthy definition of culture. I summarize it as follows:

> *Culture is the systematically developed striving toward the to-be-acquired process-moderated sum of labor by the sum of humanity. This striving envisions for itself the task of discovering all created forces, developing them according to their distinct natures, and making them serviceable for all near and far environments. This must take place in*

---

31. When I obtained my doctorate on the topic "common grace" in 1966, I made a number of critical comments in my dissertation about the concept of the "cultural mandate." In the circle of theologians and politicians who are grateful for the ideas of Klaas Schilder, the concept of the cultural mandate was (and is) a key notion. The criticism of the concept became especially timely when, shortly after 1966, the environmental problem began to attract attention. For the well-known expression in Schilder, see his *Christ and Culture*, 40.

The Bible and the Environment

accordance with the norms of God's revealed truth. Its goal is to make the results usable for the service of humanity toward God. Together with the human beings increasingly equipped for this service, the results are placed at God's feet, so that he will be all in all and all work will glorify its Master.

This definition does not contain something Schilder had posited earlier in his argument, however: the cultural mandate is maintained despite the fall. In the middle of history, Jesus Christ, as the second Adam, went back to the beginning and began the reformatory task of restoring the ABCs of life and world order. He reminds his people again of the agenda given at Creation. Schilder finds this expressed in, among other places, the parable of the talents. All the talents that God handed out to his laborers on the morning of creation have to be utilized, so that the laborers can extract from the world whatever it contains. Human beings have to discover the potentialities of Creation and respect them according to their kinds.[32]

EVALUATION OF THIS VISION

How shall we evaluate this viewpoint? I shall restrict myself to those elements that are relevant to our topic of human beings and the environment. In the first place, there is no question that Genesis 1 is speaking about our present culture. Although the culture mentioned in Genesis 1 may be primitive, the exercise of dominion and subjection, entrusted to human beings in that chapter, is indeed culture. As we saw, the subduing of the earth involves its cultivation, because human beings derive their food from it. Thus, in its most original meaning, to subdue refers to the tilling and cultivation of the land. But why should the mining described in Job 28 not amount to an extension of the subduing of Genesis 1? We read in Job 28:1–11 that human beings put their hand to the flinty rock, cut mine shafts, dammed up the underground streams,

---

32. On this aspect, see Schilder, *Christ and Culture*, 53–65.

and brought "the thing that is hidden out to light." Moreover, why should further development of technology be excluded from the subduing of Genesis 1? The primitive beginnings of agriculture have developed into the complex culture of our time.[33] A population that now numbers in the billions is apt to use the earth in a different manner than the handful of people who could survive on a primitive form of agriculture. Human dominion and subduing have developed in many different directions, such as gathering minerals, manufacturing ships and airplanes, and practicing all forms of art.

When you examine the Hebrew words for "have dominion" and "subdue," you will discover that *in the Bible* both words always refer only to animals and people, not, for example, to inanimate materials or artistic creations. But this conclusion need not prevent us from seeing mining and art as forms of the culture that began in Genesis 1. Many matters that are not mentioned in the Bible can be regarded as an extension of what is described in it.

More to the point, can we speak about culture in such a *programmatic* way, like Kuyper and Schilder did? Schilder defined culture as the to-be-acquired, process-moderated *sum* of labor by the *sum* of humanity. Kuyper wrote that the fulfillment of the world could occur *only* when everything that God had hidden in nature and the world has come to light.[34] In my opinion, these statements go too far. Do we really have to make use of every potentiality concealed in the world? The environmental crisis has taught us to express ourselves more carefully. Significant dangers may arise from the idea of "extracting everything the world contains." Misgivings about nuclear energy, nuclear weapons, and genetic engineering are proof that we must also define the *boundaries* of our cultural possibilities.

---

33. I disagree with Moltmann, *God in Creation*, 20–21, that "subduing the earth" in Genesis 1 has nothing to do with the *dominium terrae* (the dominion over the earth), and is only a commandment for the nourishment of people. Nourishment is undoubtedly the initial meaning of "subjection" (using the fruits of plants and trees the earth generates), but it is not, for that reason, the only one.

34. See my *Algemene Genade*, 372.

The Bible and the Environment

Besides, the Bible itself does not supply us with a program for our cultural task or its progressively advancing character. We cannot adduce arguments from the Bible to prove that our cultural achievements will be used as building blocks for God's kingdom. Moreover, we cannot discern a prefiguration of the renewal of nature in the continuing development of technology and medicine, as has been argued.[35] Instead, we can point to 2 Peter 3:5-11, which says that the earth will be destroyed by fire, a passage that draws a sharp distinction between the old and new worlds. Therefore we are wise to avoid speculation and stick with the simple conclusion that Genesis 1 speaks of a cultural calling,[36] but does not reveal anything about a *program* for that calling or its end result.

We can say, however, that the total subjection of everything involving humanity has been given to the man, Jesus Christ. What Psalm 8 says about the glory and honor of human beings is applied to Christ in Hebrews 2:5-9: "[Y]ou have crowned him with glory and honor, putting everything in subjection under his feet." That should also prompt us to modesty when we consider culture. Being culturally active in the broadest sense can and should be based on the Bible. But a cultural program that has to be completed by

---

35. Berkhof, in the already cited report. In Bristol, Berkhof's report was attacked especially because of its optimistic nature. Schlink, *New Directions in Faith and Order*, 91, said that the report failed adequately to underscore the New Testament's particulars about rupture and judgment.

36. The *blessing* that God gave to the first two human beings ("And God blessed them. And God said . . .") provides the strength that human beings require for the procreation, inquisitiveness, and intelligence they need to exercise dominion and cultivate creation. But that does not exclude their own *calling to work and exert themselves*. Gift and task are indivisible. Human beings are different than fish and birds because they hear the command in the blessing given to them. See my *Algemene Genade*, 377. Hence, I do not share the opinion of H. Leene, who argues that human beings receive a blessing to exercise dominion, but at the same time says that the cultural task is unfortunate. According to him, the Old Testament does not recognize "the idea of a gradual process of enculturing the earth" at all. One can argue about the concept of "gradual," but not about the fact that human beings have been commanded to subdue the earth. See Gerard Dekker et al., eds., *Werken: Zin of geen zin*, 21-22.

human beings before the history of the world can come to its conclusion would accord too much honor to our accomplishments.

Therefore, we ought to avoid the expression "extract from the world everything it contains." Besides, we have learned from the environmental crisis that it is sometimes quite necessary to leave in the world what the world contains!

My objections do not imply that accepting Schilder's definition is bound to lead to the exploitation of nature. Although Schilder was not acquainted with the environmental issue (he died in 1952), his definition contains elements that definitely *oppose* environmental destruction. For instance, he spoke about the development of creation potentialities "according to their own characteristics." When you limit yourself to that, you are doing justice to plants and animals.

We also read in Schilder's definition that creation potentialities must be used to serve near *and far* environments. This sounds similar to the "organic growth and development" Pestel described in his report, *Beyond the Limits to Growth: A Report to the Club of Rome*. Little by little, we must decide *together* what may or may not be allowed to grow in the world. Unlimited individualistic and regionalistic conduct will be detrimental for the "farthest environment." There is only one world and its management has been entrusted to humanity collectively.

Finally, Schilder rightly reminds us that we are bound by God's revealed truth, we must praise him, and we must deposit the results of our labor at his feet. That is the correct theocentric attitude that we, as good stewards, ought to adopt, as I noted earlier.

CULTURAL MANDATE AND PILGRIMAGE

Although the Bible does not outline a cultural program that must be completed before the end of history, we do read of another program, namely, that the gospel of the kingdom must be proclaimed throughout the whole world, and *then* the end will come (Matt 24:14).

In his publication *Ethiek en pelgrimage* (*Ethics and Pilgrimage*), W. H. Velema therefore wonders whether the Christian's task is to implement the original directive given to human beings at their creation in God's image, or whether the Christian's work as "pilgrims" on earth is primarily missional. Velema has no difficulty accepting Genesis 1:28 as an abiding task that remains in effect even after Christ came to earth, for the phrase "cultural mandate" does not, on its own, suggest any relationship to Christ's work. Yet our labor today is determined by history. Christians must carry out cultural work in the context of Christ's redemptive work.[37]

Indeed, it is critically important to point out the relationship of the cultural mandate to Christ. We have already seen that, as second Adam, he is Lord over all creation. That knowledge in itself keeps us from proposing a cultural program that we must complete before history can be concluded. God reaches his intended goal, also concerning flora and fauna, not through our efforts, but through Christ's work.

Yet we do not have to choose one or the other: either pursue the cultural mandate or proclaim the gospel. Suitable work in and with the world is service to Christ, who exercises dominion over the world. And so there is no difference in *principle* between the work of a minister or missionary and that of a baker, butcher, and field biologist. All of these work with and for the world. As far as that goes, we need not consider ourselves to be "pilgrims" or "strangers," because we are working in the service of our Lord.

According to the Bible, we are pilgrims *in specific respects*, as Christians among a majority who reject and are hostile toward the faith. We are pilgrims because life is difficult and we are on the way to a better, heavenly country.[38] We are visible as pilgrims when the style of our otherwise ordinary work manifests that we are Christians. And I consider our efforts on behalf of the environment as ordinary work.

The gospel does not come to us with a brand new program to which all people must devote their efforts, discontinuing their

---

37. Velema, *Ethiek en pelgrimage*, 47–58.
38. See my *Algemene genade*, 378–79.

former tasks. The gospel renews this old world along with its manifold kinds of labor. We continue to implement the mandate of Genesis 1, also, when we dedicate ourselves to protecting the environment. Thus, the cultural mandate and spreading the gospel are not two disparate tracks. The gospel demands more than verbal proclamation. It calls us to a faith that brings forth *fruits*. And why may we not observe those fruits when we are busy with technology, with art, and with protecting the environment?

The cultural mandate and pilgrimage are not contradictory alternatives. Wherever we seek to engage in culture in a Christian manner, pilgrimage is automatically involved. To become pilgrims, we really do not need to leave the world or travel through it timidly as nomads. Anyone who belongs to Christ and lives governed by the injunction, "Be holy, for I am holy" (Lev 11:44, 1 Pet 1:16), is a pilgrim. Thus, the real choice is not cultural mandate *or* pilgrimage, nor is it actually cultural mandate *and* (in addition to) pilgrimage, but cultural mandate and *thus* pilgrimage.[39]

Therefore, when we promote environmental matters *well*, we do so as pilgrims, just like when we are engaged in other endeavors. For surely it is clear by now that those who really want to protect the environment must start with themselves and must crucify their egoism. People who are serious about that and act accordingly will soon notice that they are pilgrims. Environmental destruction is simple enough to discuss, often in alarming terms, but it is much more difficult to do something about it. For that means giving up a significant portion of our prosperity while renouncing even more luxury. And who really wants to do that? Those who do are recognizable as pilgrims!

When people like Jacobi put the Bible on trial, we can remain calm. The Bible is not guilty of encouraging environmental pollution. But when we are criticized because we are interested solely in our own advantage, just like people of the "world," we must examine ourselves. We know about the cultural mandate, but do we also know about the *Christian* lifestyle—which always necessitates self-denial—within which the cultural mandate must function?

39. Thus Klapwijk, "Christus en natuur," 134–35.

Do we dare to be nonconformists and thus pilgrims, or do we placate ourselves by thinking that the environmental situation will be bearable in our lifetime at least?

CONCLUSION

When we survey what we have discussed in this chapter, we can summarize it as follows:

The Bible does not speak of the "environment" as a separate matter of concern. The *struggle* between human beings and nature is just as apparent in the Bible as the *respect* people are to have for the excellence and the beauty of their fellow creatures.

The Bible does not contain specific texts about the environment, but it does provide general guidelines that are useful for us in the current environmental crisis. These include the awareness of the unique position we occupy in creation, as God's representatives, above plants and animals. At the same time, human beings must respect other creatures "according to their kinds." People do not have the last word; God as Creator does. The vision of the Bible is neither anthropocentric nor cosmocentric, but theocentric. That is why human beings, just like all other creatures, are in God's service. Things, plants, and animals are not at people's command; rather, human beings must care for everything to the glory of God, as his stewards.

We may continue to speak confidently about our cultural mandate, as long as we avoid speculations like Kuyper's and do not interpret the mandate to permit us, programmatically, to "extract from the world what the world contains." The environmental crisis has taught us to be careful with that approach.

We must do our work in the world in service to Christ, the second Adam. Christians must devote their efforts to proclaiming the gospel, but just as much to managing the world. Cultural mandate and pilgrimage are not contradictions. Those who want to be good stewards will automatically discover that it means struggle and self-denial—things that should not surprise a pilgrim-Christian.

3

The Agenda
What Needs to Be Done?

NO EASY SOLUTIONS

WHEN WE BECOME FULLY aware of the extent of the pollution of our environment, the question "What needs to be done?" can make us despondent. Has so much been destroyed already that recovery is impossible? Is the damage reparable when so many different interests conflict? And the idea that I formulated toward the end of the previous chapter is incredibly attractive, isn't it—it will not be so bad while *we* are alive?

Indeed, the subject matter is complex. In this ethical consideration of the matter I do not have a remedy for every facet of this dreadful state of affairs—assuming that recovery is still possible. Generalized solutions for the environmental crisis will immediately look cheap because what *must* be done is not always what *can* be done. So many conflicting interests are in play. What environmental organizations would very much like to happen often cannot be done because the economic cost is too high. For instance, when you build a freeway, you have to sacrifice a portion

The Agenda

of the countryside. But if you do not build it, there will be serious traffic problems. Since there are many sides to almost every environmental issue, finding a solution is difficult. In addition, the environmental question has acquired a global dimension. This means that many authorities, all of which have their own interests, have to work together to clean up the environment. Think about the Rhine, for example, and you will appreciate how much transnational effort it takes to restore such a river.

ROMANTICISM AND COMMON SENSE

We could expect that an ethical consideration will at least identify the *attitude* we must have if we are to tackle the problems confronting us. Well then, in the previous chapters I have tried to make it clear that God has given us a mandate to become involved in the protection and improvement of the environment.

At the outset I want to make it clear that we cannot go back to the time of sailing ships, oil lamps, and exclusively low-rise buildings. We may not ignore the pattern of developmental progress in the world. Rather than idolizing the past, we must look for answers that suit the current situation. We cannot go "back to nature" (Rousseau)—a nature that was not all that peaceful and amiable. Nor can we return to small economic and technical associations. "Small is beautiful"[1] may be successful in certain sectors, but in many respects we need far-reaching centralization[2] in addition to decentralization, *precisely* in order to tackle the global envi-

---

1. This is the title of a book by Schumacher: *Small Is Beautiful*. The book describes useful elements that cannot be rejected disparagingly as "pixie technology" in this modern world. See Nijkamp, *Maatschappij*, 158ff. But at the same time we must conclude that the alternative this book presents for large-scale industrialization offers only partial solutions. Most things can no longer be kept or made small.

2. The computer, to which we can ascribe centralizing tendencies, also has remarkable decentralizing effects. I am thinking about what Toffler, *The Third Wave*, predicted about it: the computer makes it possible to do at home what would otherwise have to be done in large offices and large factories. See my *Vrede in de maatschappij*, 85–98.

ronmental issue. We must reject a romantic nostalgia for a past that cannot return. Turning back the clock would be a disaster when we consider complex economic structures and technological inventions or skills on which millions of people have become dependent.

There is progress in the life of the individual, but also in that of the world. Just as it is impossible to relive your childhood years, so too the world is moving from its beginning to its end. Our individual lives on earth come to a conclusion, but so too will the world. We read in Psalm 102:25-26, "Of old you laid the foundation of the earth, and the heavens are the work of your hands. They will perish, but you will remain; they will all wear out like a garment. You will change them like a robe, and they will pass away." We find the same in Isaiah 51:6: "The heavens vanish like smoke, the earth will wear out like a garment." This knowledge inspires us to be levelheaded. It should not surprise us that this world will become exhausted at some point.

Yet our confidence that God guides history and reaches *his* goal does not make us careless in environmental matters. But that confidence does give us peace, even if the environment should deteriorate further: "The issues of environmental pollution and of the supply of energy and of raw materials lose their deadly sting when we expect a new heaven and a new earth."[3]

IMPERMISSIBLE METHODS

When we are aware of the finiteness of both people and nature, it is easier for us to look critically at certain attempts to make the environment "ideal" again. Those who promote, for example, abortion, euthanasia, and homosexuality because these practices help curtail the world's population, are promoting improper methods. Overpopulation is a serious environmental concern, but that does

---

3. A. Klapwijk, Thesis 11, in connection with the dissertation "Eliminatie van stikstof uit afvalwater door denitrificatie ("Elimination of Nitrogen from Wastewater").

The Agenda

not entitle us to save the environment at the cost of human lives or by advocating unnatural sexual relations.

We must also reject other "realistic" solutions that conflict with Christian ethics. We should be leery when someone advises the churches to limit their humanitarian activities because relief actions lead to a net increase in the population and can thus lead to more misery. Ecologists who present such arguments are not really defending the environment, but rather their own interests. They argue that we can avoid environmental disaster if we act sensibly, without scruples. Others will perish, but we will survive. The population *must* decrease, so for a better future we really *need* a disaster.[4] A similar "lifeboat ethic" that guarantees us a place onboard but dares to consign to death all who are poor, weak, and wretched has absolutely nothing to do with a responsible environmental policy.

It is different with *birth control*, because that functions to prevent birth rather than to kill the unborn. There are currently some 5 billion people on the earth and this number is expected to grow to more than 7.7 billion by the year 2020. This will be accompanied by massive urbanization, so that Mexico City is expected to have 30 million inhabitants, Mumbai and Calcutta 20 million each, Jakarta 17 million, Cairo 16 million, and Manila 13 million. If this population explosion continues, there will be approximately 30 billion people in the world by the year 2100. However, a recent projection suggests that the world's population will stabilize at around 10.2 billion by 2085,[5] assuming that the number of births will not continue to increase at the current rate. Such stabilization will be fortuitous rather than disastrous because the more people there are, the more pollution there will be. If pollution causes almost insurmountable problems in the current industrialized world, how can we clean up the environment when the population continues to multiply?

---

4. According to Derr, *Ecologie*, 155–56, who rightly takes a strong stand against this "realism."

5. For this data, see Bibo, "Wat is er aan de hand," 31.

Environmental Stewardship

In my view, arguing that no one may ever use birth control in order to control population problems is untenable. If overpopulation really exists, and if it is a threat, then we are responsible to find a solution to that problem. By way of analogy, since lightning rods exist, we should not refuse to use them, thereby allowing our buildings to burn down, arguing that it is God's will. If you have any inkling of the problems that plague the rapidly increasing populations in countries like China and India, or in densely populated regions like Java, you will understand that governments cannot always avoid legislating guidelines for family planning. But when they demand that parents have only one child and when, as a consequence, parents abort a child or commit infanticide because they want a son rather than a daughter, then we must condemn such population politics.

We should note that overpopulation was a much greater concern in the 1970s than it is today. The world has so many food and energy sources that it can sustain a multiple of the current population. However, that means that the available resources must be used profitably and distributed honestly.

Today some politicians are actually again calling for larger rather than smaller families. This has happened, for example, in Western Europe, where otherwise the population will age and the economy will stagnate because there will be an insufficient number of young people. This example demonstrates that environmental issues cannot be separated from other factors. There are no easy answers. Some overpopulated areas of the world still have too many inhabitants for an inadequate food supply. And other areas are underpopulated because there are not enough inhabitants for the economy to function optimally. A smaller number of people does not necessarily mean that the environment remains safe. Think about the logging and erosion in Africa, for example. And a large number of people does not automatically mean that the environment will be destroyed. Some densely populated cities and districts manage to control adverse environmental consequences by means of sound environmental policy.

PERSONAL ATTITUDE

We may not try to save the environment by improper means in order to prolong the world's existence. But neither may we shorten its existence by not adopting all measures against environmental pollution that are available to us. As Christians we have a clear mandate to restore and preserve the environment. It does not help simply to lament. Nor is it sufficient if, in the current environmental crisis, we see only the judgments of God prophesied in the book of Revelation.[6] We have work to do.

The obligation to work is first of all *personal*. Certainly, our task is not complete when we merely give advice to others about their personal use of the environment. We need to work together collectively, and for that national and international decisions are necessary. But such decisions must be supported by our personal attitude toward the environment. A "cowboy economy" that follows the principle "use something once and then pitch it" is unacceptable. The consumer mentality apparent in this kind of economy is incompatible with our stewardship, as it always has been. But now that we know how wasteful we have been for decades with natural resources in wealthy countries, we reflect upon our wastefulness. Only a small proportion of the world's population shares our affluence, and the gap between rich and poor countries has meanwhile become greater rather than smaller.

The call for self-examination and environmental action has led churches to plead for a "new lifestyle." In that new lifestyle we are supposed to renounce regular increases in income, and our consumption has to become more critical and austere. We should not let ourselves be governed by the pursuit of money and goods.

---

6. Van der Poll and Stapert, *Als het water*, 48ff., provide a summary of what has been written in evangelical circles about the environmental issue. For example, Billy Graham believes that the pale horse of Revelation 6 portrays the consequences of environmental pollution. He regards pollution as the judgment of God upon the ungodly. After their analysis of Graham and others, the authors find it shocking "that, as an evangelical movement, we can, in our living, preaching, writing, and pastoring, almost completely disregard one of the greatest problems of our time, probably the greatest danger for survival on earth" (71).

Environmental Stewardship

Practices of disengagement, fasting, and abstinence should be part of our lifestyle, for the benefit of nature, the environment, and the world's poor.[7]

Such calls seem to suggest that the world's fate lies in the hands of people. They barely acknowledge (if at all) that the gospel is a message that seeks to be *believed* by rich and poor. And surely we are entitled to identify that as the first task of the church, aren't we? The church must proclaim where true riches can be found: not in a "human," earthly existence that the rich must prepare for the poor, but in the gospel of the deliverance from sin, which *is* already a fact in Jesus Christ for all who believe, whether rich or poor.[8] The foundation of the call for a new lifestyle is clearly horizontal in character. Moreover, zeal and urgency for the global environmental crisis often fades quickly. Some churches jump quickly from one current topic to the next.

But these criticisms do not permit us to disregard the call to a new lifestyle. I am convinced that the topic will impinge upon our consciousness constantly, and most likely in an increasingly urgent form. Cultural mandate and pilgrimage go hand in hand, as we saw at the conclusion of the last chapter. Well then, a Christian lifestyle calls us to deny ourselves: "Is it not at times shameful that the average well-fed person living in the West can consume an amount of protein that is seven or eight times greater than that of the average malnourished person living in India, partly because

---

7. See, e.g., *Bezitten of bezeten zijn*, a document of the general synod of the Reformed Churches in the Netherlands (Synodical) at Haarlem (1973–75).

8. Also in the *conciliar process* in preparation for the world assembly in 1990 about " justice, peace, and the wholesomeness of creation," it is remarkable that "wholesomeness" and "peace" are presented as things that must still be *achieved* though human effort. See, for example, *Een verbond voor het leven*, of the Steering Committee of the Council of Churches in the Netherlands. The document does not describe this "covenant" as a matter between God and people who believe in Jesus Christ as Savior from their sins. The covenant encompasses *all*, specifically the poor and those who are threatened (whether believers or non-believers). And it is apparent that all can devote themselves to this covenant, no matter what their faith or philosophy of life.

we in the West with our high incomes dominate the international grain market?"⁹

That is why it was correct for a synod to address the church members as follows:

> We urge families to build into their household budgets thrift, economy, and giving as large a percentage as possible for the benefit of the neighbor, both near and far. It is apparent that with respect to responsible household budgets there is a serious carelessness. A return to living simply and frugally, and with a joyful readiness to share, is very necessary. In order to break through the current misplaced developments, it is urgently necessary that we reflect constantly on giving expression to a unique Christian consumer style in these times.[10]

Sometimes we are frightened when confronted by the most severe suffering in the world. Regrettably, these times are often just *moments* in our typically superficial life. We have brought nothing into the world and cannot take anything out of it, as we read in Scripture (1 Tim 6:7). When this simple truth leads to the lifestyle that the Bible commends, it will have a positive effect on the environment.

By our personal efforts, we can easily benefit the environment by abandoning our above-mentioned carelessness. For example, we can all help by choosing phosphate-free laundry detergents, avoiding the use of aerosol cans, plastic wrappers, and disposable bottles and cans, and by separating different kinds of garbage, and so on.[11] We can justifiably criticize others about environmental pollution only after we acknowledge and strive to overcome our own culpability.

---

9. Nijkamp, *Maatschappij*, 143–44.
10. *Bezitten of bezeten zijn*, 24.
11. Nijkamp and Douma, *Het gelaat*, 62–63.

Environmental Stewardship

POLITICS

Our personal attitude is important, but by itself is inadequate. Speaking about personal abstinence amounts to no more than boastful prattle if we rich people consume a little bit less, pass up a few luxuries, and then think that we can thereby solve the problems of the world. Only the naïve think that if each of us uses fewer resources, and if the hungry of the Third World decide not to strive for a consumer society, then we have already achieved much good. On the global level, wealthy countries will have to make sacrifices that really hurt. Undoubtedly, we will not end up with a better environment unless we also strive for greater economic justice.[12]

And that brings us into *political* territory on a national and global level, for governments are obliged to maintain sound, visionary environmental policies.

Such policies involve many factors. Take, for example, people's employment and housing conditions. Even if the government does not have to take measures to curtail population growth, it does have the task to *decentralize* the population. Urbanization, which is always deleterious for the environment, has to be regulated properly, maintaining a good balance between opportunities for employment, housing, and relaxation.

A sound environmental policy also makes provision for the countryside. The tons of garbage dumped there over the years must now be cleaned up. Additionally, governments should implement recycling programs and should require manufactured products that have a longer life. We must also strive to curtail farming and animal husbandry, which are the source of an irresponsibly high degree of water pollution and acid rain. Alternatively, we must pursue methods of farming and animal husbandry that result in less pollution.

Heavier penalties are necessary to deter environmental pollution by industries and individuals. Since 1970, stringent government policies have permitted a polluted city like Singapore to be cleaned up. All littering—even of orange peels—was and is heavily

---

12. Derr, *Ecologie*, 152–53.

penalized. Citizens were encouraged to plant flowers and shrubs in their yards, and flowerbeds were planted in traffic circles and street medians.[13] Clearly, a different mentality can do wonders for the environment, but that will require encouragement and correction by the government.

Our continually increasing automobile usage also causes large problems. Improvements to internal combustion engines and exhaust systems and the mandated use of lead-free gasoline can reduce pollution, but if parking lots continue to grow larger the gain from such measures is too limited.

Large expenditures and strict regulations to clean up cities and the countryside and then to maintain them in that state will, of course, take their toll on other matters. Environmental management demands self-discipline. Thus, we ought to make modest use of private enjoyments, such as our cars.

Those who pursue their own interests and their immediate self-gratification will not be all that concerned about the environment. For economic products are often more expensive to produce when you have to observe the rules of sound environmental management. Environmental abuse is typically profitable in the short term, but do not ask what it will cost in the long term. Indeed, that is why our attitude must change. We must take all interests into account in order to reach and maintain an optimal environment.[14]

We must learn to consider the long-term effect of our actions and make decisions accordingly. Increasingly, people will have to replace their desire for immediate gratification with the perseverance that characterized builders of cathedrals and designers of baroque gardens in the seventeenth century. Pestel uses this comparison: the builders and designers knew that they would not live long enough to see the completion of their work, but they began and continued it.[15] Environmental policies must display a similar vision. Moreover, we must be prepared to make the sacrifices

---

13. According to a news item in *Trouw*, February 27, 1975.
14. Van den Berg et al., *Bouwen en bewaren*, 42.
15. Mesarovic and Pestel, *Turning Point,* 59.

necessary in the short term in order to guarantee a livable environment for our children and grandchildren.

A national environmental policy is insufficient by itself, however. Environmental damage does not stop at any country's border and, therefore, it requires *international* attention and action. Think, for example, of the pollution of the air, oceans, and large rivers. Salt from French potash mines pollutes the Rhine, which then flows through Germany and the Netherlands. The forests of Scandinavia and Central Europe are being poisoned partly by smoke and exhaust gases from England. Unfortunately, each country is inclined to pursue its own advantage, even at the expense of others.

There is no common authority that can impose measures like those legislated in individual countries. Nevertheless, nations will have to reach bilateral and multilateral agreements to prevent disasters that can affect them all. Indeed, such agreements are a global need, although we should not strive for a one-world government.[16]

If anywhere, then certainly in the context of the environmental crisis, economic factors play a dominant role. A balanced environmental policy of global dimensions is impossible without tackling the huge disparities that exist between the rich and poor countries of the world. We shall have to discuss the most pressing environmental problems in the context of an improved world economy.[17]

You can easily become discouraged when thinking about all the changes necessary to achieve a clean environment worldwide. We must acknowledge, therefore, the results achieved by sound environmental policy, no matter how limited the extent of the improvements so far. The air in the Ruhr district is now cleaner than it was fifty years ago and the water in the Swiss and Bavarian

16. Tinbergen, *De aarde beheren*, 162ff., points in this direction. Despite all the criticism that can be leveled against such a vision (a world federation with a strong centralized authority is utopian, dangerous, undemocratic, etc.), it should be clear that global guidelines for the protection of our shared environment will become necessary and thus can no longer be regarded as utopian.

17. See, e.g., Von Weiszäcker, *De tijd dringt*, 106.

The Agenda

lakes has similarly improved in the last two decades.[18] We could list other encouraging improvements as well. Serious concern is appropriate, and we may ask ourselves if we still have enough time to avert the crisis. But overwhelming concern should not cause us to slacken our efforts. In the meantime there have been successes that can spur us on. The protests of environmental groups have not been in vain, since almost every country is now paying attention to the environment at the highest governmental levels, and environmental initiatives are underway internationally as well.

THE PLACE OF THE ANIMAL

I want to devote more extensive attention to two topics. First I will discuss the protection of *animals* in our struggle for a better environment. If we do not do everything possible to restore the environment, animals will suffer more than they do already. Our special relationship to animals justifies a separate discussion about the protection of animals.

I have already described the multiplicity of life forms, to which God gave each its own nature. Accordingly, we must respect and interact with animals as God's creatures. Yet, does this mean that animals have "rights," just like human beings? Must all animals be protected because they, like human beings, have a right to life?

Such arguments are indeed being defended with varying degrees of intensity. According to Peter Singer, for instance, *speciesism* is wrong, akin to racism. By this he means discrimination against species. The species of human beings, monkeys, whales, pigs, and mice are alike in one cardinal aspect: they can suffer pain and experience desire. The determinative question is not whether animals can think or speak but whether they can suffer pain. That is why there is no sound basis to protect human life but not to protect animals. Singer detects the consequences of speciesism

---

18. Ibid., 47.

in, among other things, consuming animal meats, biotechnology, vivisection, and hunting animals.[19]

Singer does qualify his position somewhat. Because the term "animal" encompasses such a wide variety of different living beings, answers to the question raised above will vary. Singer therefore draws a distinction between higher and lower forms of animal life, with the higher forms having a kind of "power of reason." According to Singer, chimpanzees, whales, and dolphins have the ability to think and to be aware of themselves as distinct beings with a past and a future. Killing such animals is therefore just as serious as killing a disabled human being who lives at a similarly low mental level. Singer believes that the list of animals deserving protection can be expanded to include other kinds of apes, dogs, cats, pigs, seals, and bears, since they also have well-developed mental capacities. However, it is permissible to kill and eat animals that can feel pain and have consciousness but do not have *self-awareness*.[20] Therefore Singer's position is not strictly vegetarian.

James Rachels expresses similar ideas. He too does not draw a principal distinction between humans and other life forms. Instead, he seeks to define whether the subject *has life* or is merely *alive*. A person who develops Alzheimer's disease, which attacks the brain and diminishes human personality, *is* indeed alive, but does not *have* a life. To have a life means to have plans, expectations, aspirations, and the ability to look back on the past and ahead to the future. Rachels also describes this as life in a *biographical* sense. But to be merely alive is something far less significant. A person who is severely mentally disabled, a patient in a coma, or a seriously demented person is alive biologically, but not (or no longer) alive biographically. Well then, Rachels proposes, some animals have a higher form of life than such subnormal humans who lack diverse valuable elements. Rhesus monkeys, for example, live together in social groups, have families, communicate with each other, and enjoy a high degree of individualized personalities. Rachels introduces a sliding scale of values. A Rhesus monkey

19. Singer, *Practical Ethics*, 47ff.
20. Ibid., 103–4.

has more life than simpler animals, such as mussels, snakes, and insects, for which life is not so valuable. That is why we find it easier to kill an insect than a Rhesus monkey.[21]

I cannot accept the positions of Singer and Rachels, although not necessarily because of their arguments about the power of reason, self-awareness, forms of communication, etc., which are no doubt present in higher kinds of animals. Yet even if what Singer posits about this (in conjunction with biological research) is true, we must still dispute his conclusions. For the value of life is not determined by perceptions such as pain and desire, or by certain capabilities such as self-awareness. The *qualities* of human beings and animals are not determinative, but rather their *place and task*, assigned to them by God. God placed human beings above every species of animal. After the flood, he also gave human beings the right to use animals for food, which is not a "discrimination against species." Discrimination always results, impermissibly, in victims. But if human beings use animals for food, this is a result of God's appointed order and (therefore) not discrimination.

We should not replace the distance between human beings and animals with a sliding scale of values. Even the most severely disabled human life exceeds that of the most highly developed animal life, because it is *human* life.

When we seek to protect certain species of animals from being hunted, we do not do so because they have an inherent right to such protection. Those animals are entitled to protection not because they are holy, but because such animals have become scarce. When some species are threatened with extinction, we must counteract that.

Moreover, we are right to distinguish between useful and harmful animals. We try to exterminate rats, mice, and all kinds of insects because they cause us trouble. In addition, we are entitled to use some species of animals for food. They are bred for their milk, meat, and egg production. If we need their meat, they will be slaughtered, for the Bible does not teach us to be vegetarians.

---

21. See Rachels, *End of Life*, 23–36.

RESPECT FOR ANIMALS

Significantly, God did not give permission to use animals as food immediately at the time of creation. In addition, the warning of Genesis 9:4 to abstain from the *blood* of animals is significant. The blood was regarded as the locus of life (the "soul"), which belongs to God, so the blood of animals killed in the hunt had to be poured out and covered with earth. People were not allowed to consume it; it could be used only for sacrifices to God (Lev 17:10–14). They were permitted to kill animals and eat their meat, but the Israelites had to abstain from the blood as the *life* that belongs to God.[22]

We may include animals in the "eager longing" of "the whole" creation, which is groaning to be set free from its bondage of corruption (Rom 8:19–23). Indeed, the suffering of animals is part of the groaning of creation. As Karl Barth noted, this passage from Romans 8 might well be posted in flaming letters above every hunting lodge, every slaughterhouse, and every vivisection lab.[23]

We may use plants and animals for our food, but only if we do not destroy vegetation in the process, but instead make meaningful use of its abundance. In that case we are not abusing the flora. And we should treat the fauna just as carefully. When we eat meat that comes from regulated slaughterhouses, or from a hunt that operates within established limits and maintains the fauna at an optimal level, we not need feel guilty.

Furthermore, we may not be indifferent to the fact that an animal suffers pain. Spinoza maintained that the sensations of an animal differ from those of people. He did not deny that an animal has feeling, but at the same time he believed that people are free to use animals as they please, because by nature they are not equal with us.[24] On the basis that animals are different, some people treat

---

22. The ban on eating blood was repeated in Acts 15:20 and was respected in the ancient church, as we know from the testimony of a martyr from Lyon. See Aalders, *Wereldkerk*, 106. For more discussion on the question of whether it is lawful to consume blood, see my work *The Ten Commandments*, 208.

23. Barth, *Church Dogmatics*, vol. III/4, 355.

24. Spinoza, *Ethics* IV, proposition 37, comment 1.

The Agenda

animals arbitrarily or even cruelly. This betrays an anthropocentric attitude, which misunderstands that people are merely stewards.

Animals are certainly different than human beings also in how they experience pain. But it is inhumane to be unaffected by the painful cries of animals. The righteous have regard for the lives of their animals (Prov 12:10) and they do not muzzle their ox when it is treading out the grain (Deut 25:4). They know when an animal is hungry and recognize also when it suffers pain. If they can prevent or minimize that pain, they should do so. Killing an animal is not murder, but when we kill it we must not cause it unnecessary and excessive suffering.

When we make an animal suffer pain, whether our actions are cruel depends on our underlying motive. People who neglect or torture their pet are guilty of animal cruelty. But we ought not to identify a person who causes animals suffering in the context of, for example, cancer research as someone who is mistreating animals. *Animal experiments* are permissible because they provide useful knowledge to medical and biological science. Human beings may not be used as guinea pigs; animals may.

It is surely a strange phenomenon that we object to the *vivisection* of (white) rats and mice, while at the same time we are totally indifferent to killing such animals in nature with traps and poison. However, it is appropriate that animal experimentation is regulated to prevent the mistreatment or unnecessary use of laboratory animals. Further, animal experimentation for cosmetic purposes is difficult to justify.[25] We have plenty of cosmetics already, so we do not need to sacrifice any animals for that purpose.[26]

---

25. Approximately 80 percent of vertebrate laboratory animals in 1981 consisted of mice and rats. Approximately 0.08 percent of those animals were intended for use in cosmetic experimentation. See the entry in *Grote Winkler Prins Encyclopedie*, 23, 245.

26. *Editor's note*: The Catechism of the Catholic Church devotes several paragraphs to the matter of respect for the integrity of creation, including animals. See paragraphs 2415–2418, which acknowledge the permissibility of using animals for food, clothing, domestication, and medical and scientific experimentation. Online: http://www.vatican.va/archive/ccc_css/archive/catechism/p3s2c2a7.htm.

Environmental Stewardship

Ritual slaughter by Jews and Muslims, employed in their sacrificial festivities, has also been criticized. Yet this criticism is usually unjustified in the context of the fair treatment of animals. Kosher slaughter for Jews is done professionally and ensures that the animal suffers as briefly as possible. Islamic slaughter often causes more suffering for the animal because the slaughter is not always done by a professional and sometimes occurs in places unsuitable for that purpose.[27]

We must also carefully weigh the criticism leveled against certain methods employed in the *animal agriculture industry* (agribusiness). Is it responsible to raise animals such as pigs and chickens in surroundings totally unlike the natural environment in which they lived before the emergence of this industry? Often the animals are cooped up together in large numbers. If pigs bite each other's tails out of boredom, then those in the animal agriculture industry simply dock their tails. They also cut the canine teeth of nursing piglets (without anesthetic) to prevent damage to the sows' udders. Castration is commonplace to prevent undesirable hormonal development in boars because their meat can begin to smell after slaughter. Even though this odor arises in only a small minority of boars, for the sake of convenience as many boars as possible are castrated.[28] What does all this have to do with the well-being of the animal itself?

We should not become sentimental in our criticism, however. For example, animals that are housed in a very small space are not necessarily under stress. A canary in a cage is also living in a small space, but it nonetheless sings. But the aforementioned agribusiness practices no longer show any or hardly any respect for animals. Producers certainly show no respect for calves when they raise them to be anemic, yielding white meat for export. Similarly, geese are overfed to obtain enlarged livers. In such cases, people

---

27. I draw this information from articles about ritual slaughter in *Nederlands Dagblad*, December 10, 1977, and in *Trouw*, January 10, 1975, and December 6, 1980.

28. Information in *Elseviers Weekblad*, October 29, 1988, 95.

have regard only for their own interests, not for the well-being of the animals in their care.

Some animals are also kept as *pets*. In such cases, they become a very integral component of people's environments. Keeping animals as pets is a very ancient practice, which means that pets are not an indication of an affluent, indulgent society. Pekingese dogs, for example, were kept as pets in the Chinese empire more than five thousand years ago, and Maltese dogs were known to be lapdogs of ladies-in-waiting during the Roman Empire. There is a tendency in our Western civilization to think of a pet as a kind of person. Pet cemeteries and lap dogs are expressions of this.[29] To be sure, attachment to a pet is a good thing, and many people cry when their dog or cat dies. But their tears may dry quickly, because the faithful companion was "just" an animal, after all.

We may also observe animals and their habits in zoos for our education and amusement. But there is a kind of amusement beneath the dignity of animals.[30] This includes "degrading acrobatics in circuses with apes, elephants, lions, bears, etc., which are forced to perform inane imitations of what human beings can do as normal expressions."[31]

We must treat animals with serious respect, as God's special creations. Animals are not human beings, but neither are they merely things. Indeed, they may not be our brothers, but they are our companions. That is why our commitment to the environment must always include an effort to safeguard animals, which have received their place in the world from God. The world is bigger than a zoo. In environmental cleanup and preservation, we often focus on achieving cleaner, breathable air, but that should not be our only goal. For, when you commit yourself to God's creation, you will also consider the animals.

29. Jeuken, *Ethiek*, 46.

30. *Editor's note:* Although unmentioned by the author in the Dutch original, surely the reference that follows, to degrading acrobatics in circuses, would apply as well to reflection on the moral implications of rodeos and bullfights, along with any other kinds of animal fighting for the purpose of human pleasure or profit.

31. So formulated by Smelik, *De ethiek*, 154.

Environmental Stewardship

NUCLEAR ENERGY

A second topic that must be discussed in greater detail is whether *nuclear energy* is acceptable as a source of electricity. Beginning in June 1981, in fact, a "broad social discussion" (Dutch acronym: BMD) was organized in the Netherlands. Of all the organizations and people that submitted detailed opinions about energy regulation, about half expressed concern about the quality of the environment or an aspect of it. On the matter of nuclear energy and nuclear power stations, 30 percent of the Dutch people opined that the environmental problem was more important that the aspects of safety, cost, and energy dependence.[32]

At first blush these results seem strange. For producing nuclear energy is better for the environment than producing energy from coal, oil, and natural gas. Fossil fuels yield only a small amount of energy, by weight or volume, compared to uranium (the raw material for nuclear energy). As a result, the waste products from fossil fuels are also *much greater* than from uranium. Astonishingly, the amount of active waste produced for one person who uses nuclear energy for an entire lifetime amounts to approximately 61.02 cubic inches. A coal-fired generating station generates much more pollution through such byproducts as slurry, poisonous metals that can poison the groundwater, and carbon dioxide and sulfur, which are emitted into the atmosphere.[33]

However, there is a "but." Accidents can happen at nuclear power plants, as we saw in Harrisburg and Chernobyl. Moreover, while the amount of waste produced by nuclear power plants is relatively negligible and therefore ought to keep the environment much cleaner, the byproducts present a particular problem because of their *radioactivity*. The waste generated by nuclear fission is very dangerous and requires a safe storage location. Different places have been considered: in the ground (salt mines, granite, clay, or in lava lakes), in water (deep in the sea), or even in space

---

32. See *Het Tussenrapport*, 122.
33. Mink, *Kernenergie in opspraak*, 107.

The Agenda

(sent up by rocket). The most realistic solution is underground storage in salt domes at a depth of at least several hundred yards.

Technically, radioactive waste can be contained. Additionally, nuclear power stations can be constructed to be so safe that industrial accidents are almost impossible. The devastating accident in Chernobyl occurred in an older type of nuclear reactor. Besides, the accident had fewer dramatic consequences than were expected initially. Generally speaking, the chance of death from a nuclear accident is very small—much smaller than, for example, the chance of being hit by lightning or of going down in an airplane crash. Comparatively, the Delta Plan in the Netherlands was implemented in order to prevent a flood more than once every ten thousand years. When such an accident happens, it is likely to cost many lives, as was the case in the disastrous flood of 1953. And yet we assume that very small risk. Scientists claim that nuclear energy is much safer than our Delta works.[34]

Economically, it makes sense to switch over (partially) to nuclear energy. The supply of coal and oil is becoming scarcer, and some reserves are exceptionally difficult and expensive to extract. For example, the Netherlands has enormous stocks of coal underground, but only 4 percent of it is recoverable by conventional mining methods. In 1974 we therefore discontinued the national mining industry because natural gas was much cheaper (and cleaner) and readily available.[35] If natural gas also becomes exhausted we shall have to turn to a new source. Hydro, wind, and solar energy (mills, turbines, solar panels, etc.) are available, probably to a much greater extent than we see now. But these energy sources are unlikely to satisfy the huge and continuous demand for energy. So why then should we not switch over to more nuclear energy?

One day the process of nuclear *fusion* may replace that of nuclear *fission*. Nuclear fusion uses the heavy isotopes of hydrogen, which is available in gigantic amounts on the earth. This technology may have many applications, possibly even as fuel for cars and

34. Van Loon, *Kernenergie*, 132–33.
35. Ibid., 23.

other forms of transportation. Furthermore, coal will again become useful and attractive as a source of hydrogen gas. Certainly this would be an interesting development: nuclear fusion would make possible a more efficient use of fossil fuels.[36]

After we consider all this, what insurmountable difficulty can we then have with nuclear energy? I suspect that nuclear energy would not now be under such suspicion had it not first been used for nuclear weapons, which led to an insecure world. Even the phrase "nuclear energy" makes many people tremble. Andrei Sakharov said this about it: "It is difficult to explain to a lay person that a nuclear reactor is not a nuclear bomb, that a coal mine or an oil drilling rig can constitute a greater danger to the environment and the health of people than a nuclear power station of equivalent capacity, or a breeder reactor."[37]

When people think about nuclear power, they are concerned first for the safety of the world, not the environment. If people were to live in sustained peace with each other, it is likely that there would be little concern about nuclear energy, even with the risk of a nuclear meltdown, for the environmental advantages are great and the risk of an accident very small.

People—not the environment—constitute the real problem. We do not live in a world of peace, so when it concerns nuclear energy we do not fear an accident so much as the misuse of, and blackmail with, radioactive material. Plutonium is made from uranium and, along with highly enriched uranium, is used to manufacture nuclear weapons. Nuclear energy, therefore, can lead to the proliferation of nuclear weapons materials. Then, through the *proliferation* of nuclear materials, more nations may acquire nuclear weapons. And what happens when radioactive material is stolen in order to blackmail groups of people, or perhaps an entire nation? Such misgivings cause many to reject nuclear energy, regardless of the peacefulness of the energy itself and of the advantages it offers the environment.

---

36. Ibid., 43.
37. Quoted in *Elseviers Magazine*, January 8, 1983, 107.

ONCE MORE: COMMON SENSE

Earlier in this chapter I pointed out that we cannot return to the time of the sailing ship and the oil lamp. That means, too, that we must dare to take risks. An increase in scale means that if something goes wrong accidents may claim more victims. For example, traffic accidents occurred during the time of the horse and buggy, but motorized traffic means that now more accidents are likely to be fatal. Everything goes faster nowadays, travel is much more intensive, and therefore risks have increased. Busses crash into ravines, traffic piles up in fog, and the resulting havoc (including the number of victims) is many times greater than when we still used the stagecoach. People like James Watt and Michael Faraday rendered the world an enormous service with their discoveries and inventions, but . . . machines can cause hellish disasters.

The electric light, the car, and the airplane are but feeble examples compared to technologies like the computer and nuclear energy. The computer is a wonderful invention, but when all our personal information is centralized, it is also a wonderful tool for a dictator. Everything that the German occupiers learned about the Dutch during the Second World War pales in comparison to such information.

Nuclear energy is an important source of power that we must accept, because minerals are running out. Wind energy may allow us to supply electricity to a house or small factory here and there, but it will not enable us to sustain the world's economy. Still, we cannot deny that one accident at a nuclear plant (Chernobyl) was enough to cause an uproar not only in a district and a country, but throughout the whole world.

The more our technology advances, the more devastating the accidents that can happen. Knowing this, we could conclude that we should get rid of computers and nuclear power stations, because the number of victims caused by their malfunction will be staggering. We need to be assured that such huge accidents are impossible before we will consent to use that technology.

Environmental Stewardship

The possibility of accidents always remains, and we overestimate human science if we imagine that people can create a society without risk. Besides, no one wants to go back to the towing barge and the oil lamp. We simply cannot return to a world without computers and nuclear fission.

Of course, we must do everything possible, technologically and politically, to prevent disasters. I have already noted that the chance of a major catastrophe is exceedingly small, but we can never rule it out completely. Even if a disaster is not caused by human error, a gigantic meteorite could fall onto an underground storage dump of nuclear waste. It has been said, correctly, that such an accident would be much more calamitous than a meltdown at a nuclear power plant.[38] Actually, the same is true if a nuclear bomb were launched against a nuclear power station or an underground nuclear storage dump.

God has permitted human beings to unlock the potentialities that he has embedded in his creation. They would be able to do that safely if they understood their responsibility toward the Creator. Sadly, we know that is usually not the case, and so our hearts are in our throats. But people often misunderstand or ignore their relationship with God. We cannot forestall disaster by seeking to halt technological development. History continues, people remain people, technology progresses, and the risks increase. But the risks are not caused by nuclear fission, just as they were not caused by the incandescent light bulb, the car, or the airplane. The historical progress of *fallen* humanity increases the chance that even greater technological discoveries will be misused, for the dizzying heights and terrible depths of the human heart are not that far apart.

Those who want to issue a warning should do so not against nuclear power plants, but against people who are estranged from God and no longer understand their responsibilities as stewards, whether that stewardship involves energy from coal or nuclear fission. A more economical energy policy, fewer nuclear power plants, and more energy from other sources can constitute sound governance. But that is a matter of governance and not a principled

38. Van Loon, *Kernenergie*, 116.

objection to nuclear energy. We can condemn and strive to abolish specific applications of nuclear energy, such as nuclear weapons.[39] But we don't have to close nuclear power plants for that purpose, and certainly Christians should not make that an issue of principle. We ought not to resist developments on the basis that they are accompanied by great risks. We are traveling through a history of great blessings and disasters. But we will survive because God is in control, not people. This may seem like a pious conclusion to this chapter, but it underscores that Christian common sense rests on a solid foundation.

---

39. Such a pronouncement does have to be made carefully, however, as I explained in my *Gewapende vrede*, 112–40.

4

Genetic Engineering (1)
Plants and Animals

A BARRIER OVERCOME

WITH THEIR INGENUITY, PEOPLE have successfully explored the tiniest particles of matter. They have been able to achieve nuclear fission and (to some extent at least) nuclear fusion. By means of these discoveries, they are able to tap into a new inexhaustible source of energy as the supplies of coal and natural gas are exhausted. But at the same time, we are aware of the drawbacks of these new discoveries. People can also manufacture nuclear weapons from nuclear material. We all realize what a disaster it would mean (also) for the environment if such weapons were used in war. Nor is the peaceful use of nuclear power without drawbacks, because it is necessary to store dangerous radioactive waste safely and indefinitely.

We discussed these concerns at the end of the previous chapter. In this chapter, I want to draw attention to another development that will probably have an even greater significance for the environment. Just as physicists have unearthed the secrets of

Genetic Engineering (1)

atoms, biologists are busy charting the smallest building blocks of living cells, which will open the way for the alteration of genetic material in plants, animals, and people. Today it is really no longer possible to discuss the environment properly unless we also consider this development.

What follows makes clear that we are dealing with an exceptional situation: all kinds of changes happen spontaneously in nature. Apple trees sometimes produce anomalous fruits, for example, and among red-flowering plants a purple one occasionally blooms. Something new has come about that preserves its anomalous form or color as a mutation (change). All kinds of external factors can cooperate to cause such spontaneously arising hereditary mutations.

People as well have always contributed to alterations in flora and fauna. By means of grafting, budding, cuttings, and the like, they have altered plants. From the wild species, they selected the plants with the largest and fattest ears, so that they would yield new and better strains. They were already engaged in biotechnology centuries ago when they made beer from grain with the help of active yeast cells. By means of selecting and crossbreeding different strains, they raised animals that produced more milk, meat, and wool.

People themselves do not remain unaltered. For example, whenever they choose to marry, they make a decision that is significant for the genetic material of their descendants. People's carelessness can also have a negative effect on hereditary material. That can occur, for example, through radiation during an accident at a nuclear power plant.

The creation, as it came forth from God's hand, is therefore not a static entity, but constantly changes. We can speak of an evolution that occurs on the foundation of creation.[1] Today's individual is a person like Adam, but in their genetic definition individuals are markedly different as a result of many mutations (think, for example, about our skin color).

---

1. This expression comes from Schilder, *Christ and Culture*, 39.

Environmental Stewardship

Evolution generally presupposes a development to a progressively higher level. But certain processes could move in a descending direction, rather than an evolutionary one, so that the already damaged creation is deformed further. That will become apparent in the rest of this chapter. We can present something as though it were development when it is really degradation.

Until recently, there was a barrier preventing us from further interventions in the genetic identity of plants, animals, and human beings. People could achieve better strains through crossbreeding, but they remained bound by the fixed genetic patterns of the relevant strains and species. They knew that these genetic patterns depended entirely upon the chromosomes and the genes in them. But this knowledge of their own hereditary factors and those of other living organisms did not enable them to alter the genetic patterns. In their experiments, scientists remained bound by the existing properties of living creatures. They were not yet able to intervene in the hereditary material of the cell itself.

Meanwhile, that barrier has been overcome too. In 1953 James D. Watson and Francis H. C. Crick, two American researchers at the University of Cambridge, England, discovered the structure of so-called DNA (deoxyribonucleic acid). This DNA is responsible for transmitting all hereditary information in plants, animals, and people. In 1869, the existence of DNA had already been noted by the Swiss researcher Friedrich Miescher, but he had no idea about the function and importance of his discovery. It would be almost another century before the fantastic significance of that discovery was understood.

To gain some understanding of the discovery,[2] we have to realize that the chromosomes of plants, animals, and human beings are composed of two DNA strands, joined together to form

---

2. For the biological-technological information I have used, see, e.g., Hermans et al., eds., *DNA-onderzoek*; Reiter and Theile, *Genetik und Moral*, 13ff.; Commission of Inquiry of the German Bundestag, *Chancen und Risken*, 7ff. This extensive report gives an excellent overview of almost all relevant aspects of the topic. See also Schellekens and Visser, *De genetische manipulatie*; Antébi and Fishlock, eds., *Biotechnology*; Eibach, *Gentechnik*, 11ff.; Seldenrijk, *Genetische technieken*.

a double helix. A large number of each of the rungs in the double helix together form one gene, which is the bearer of a hereditary characteristic. Although scientists knew for a long time about the existence of chromosomes, and were also aware that these chromosomes are the warehouses where genes are stored, since Watson and Crick's discovery they have known the structure of genes.

Under normal circumstances, people have 46 chromosomes in the nuclei of their cells. The bearers of our hereditary predisposition, genes, are found in those chromosomes. They number between 50,000 and 100,000. Whether we will have black or red hair, brown or grey eyes, small or large stature, etc., is determined from the earliest beginning of our lives by these genes. The order and composition of the rungs in the DNA double helix are determinative for an endless variation and combination of hereditary factors.

To give an impression of the enormous complexity of a small human cell, I need only point out that the extremely fine DNA double helix in a cell has three billion rungs. A small error in the construction of even one rung can cause a serious hereditary disease.

The mapping of DNA may result in establishing the exact composition and location of perhaps all hereditary factors within a living cell. Even more important, by means of *recombinant DNA technology*, it has become possible to intervene in the hereditary structure itself. Scientists have discovered substances that permit, in a manner of speaking, DNA to be cut into pieces. And then they are able to combine pieces of different origin.

If we give our imagination free rein, we can easily envision all kinds of hereditary characteristics being moved from one species to another by means of this "cut and paste" process. This is called *genetic engineering*, by which is meant the purposeful alteration of hereditary material.

New DNA combinations that do not exist in nature are now possible. In 1984 the front page of the magazine *Nature*, in which Watson and Crick had announced their momentous discovery to the world years earlier, displayed a super-mouse. In a laboratory, scientists had bred a mouse in which they had inserted a gene from a rat. This gene was responsible for the development of the growth hormone of the rat. Transferred to the mouse, this gene caused the mouse to grow like a rat and become twice the size of a normal mouse.[3]

It is possible that the super-mouse will disappear from the scene. But what will not disappear is the new skill to fiddle with genetic material. The path is now open for people to manipulate plants, animals, and even themselves. In the past they could do that up to a point, but now they no longer have to stop at what used to be genetically fixed.

For what is still fixed? Does it not seem that we can create all kinds of new living organisms? Will there come a time when qualities of plants can be incorporated into human beings? The boundaries are becoming blurred. If super-mice are possible, then in principle why not also super human beings, with new qualities that could raise people above the irksome impediments to which they are still subject? We can redo creation, and do a better job, as someone has said. We can do a better job than God.[4]

ACCEPTABLE APPLICATIONS

Before we weep and wail about all this, we should first realize that many among us are already benefiting from what this newly gained biological knowledge has produced. I am thinking in the first place about various *medicines* that are now available. I shall mention a few of them in the next chapter.

But new techniques can also be valuable for *cleaning up the environment*. Engineered bacteria that break down matter that is foreign to the environment are already being used for biological

---

3. In addition to this mouse, a hybrid of a sheep and a goat has also been in the news. This animal was created by the faculty of veterinary medicine at the University of Cambridge in 1982. However, it was the fruit of the combination of embryonic cells, not of DNA technology.

4. *Beter dan God* was the title of a three-part television program broadcast by the VPRO (Vrijzinnig Protestantse Radio Omroep [Liberal Protestant Radio Broadcasting]) in March 1987, in which the new developments were examined critically. A fourth part was broadcast a year later in March 1988.

Genetic Engineering (1)

scrubbing. In 1980 an interesting case involving patenting a bacterium was argued before the United States Supreme Court. A researcher at the University of Illinois, Ananda Chakrabarty, had developed a microorganism that was able to consume oil. This new bacterium could ingest oil from polluted areas. In addition, it actually found the more poisonous components of the oil very tasty. No wonder that the "inventor" of this bacterium recognized the economic value of it and therefore applied for a patent. Who would deny the benefits of "his" engineered bacterium for cleaning up beaches polluted with oil, for example?[5]

In addition, I would point to developments in *agriculture*. Biotechnologists are looking for opportunities to make plants resistant to diseases, frost, drought, and herbicides, to increase their tolerance of salts and minerals, and to have them produce poison against harmful insects. Scientists are also attempting to improve the protein value of edible plants, and to transfer nitrogen-bonding qualities of certain bacteria into edible plants. In that way it becomes possible that the plants themselves can derive nitrogen from the air. Such plants will then no longer require the addition of nitrogen in the form of fertilizer. They provide their own fertilizer and, as a result, a higher yield at a reduced cost is possible.

We have learned of a more modest application of DNA knowledge in the judicial context. A case that attracted general attention in the Netherlands involved a man who had been charged with rape but was acquitted because laboratory tests disclosed that his DNA was not the same as the DNA discovered in the evidence of the crime. The man, who was strongly suspected of the crime, would probably have been convicted if DNA testing had not been possible. Also, when a baby was spirited out of a hospital in Dordrecht and later recovered, DNA evidence clearly established that it was the same baby.

---

5. See Antébi and Fishlock, *Biotechnology*, 187. In fact, this bacterium was not the product of recombinant DNA technology, but of crossbreeding four bacterial strains.

Environmental Stewardship

DOES ETHICS LAG BEHIND?

We are being pressured from two sides. Genetic engineering or genetic manipulation can be a good thing or a bad thing.[6] The word "manipulation" does not sound good, but "manipulating" can mean simply "handling." It derives from the Latin word *manus*, which means "hand." *Manipulus* means "a handful." Manipulating can acquire the secondary meaning of working craftily to achieve a desired goal. In itself, that is not a bad thing either. The question is, what *goal* do we want to achieve? Whether we speak of "genetic manipulation" or "genetic engineering," we are speaking of the goal-oriented handling of genetic material.

So then, what is this goal? When the ingenuity of many talented researchers focuses on a therapeutic goal, they are using their knowledge in a good way. But when people fear that genetic engineering can also be used for an improper purpose, so that the goal of the manipulation is totally unacceptable, such fear is certainly not baseless. Does creation have to be improved? Should we not be concerned about the differences that exist between plants, animals, and human beings? May we regard the world as a Lego toy, whose building blocks we can use either for making something or tear things down, as we wish?

The developments are rapid. It has been said that the distance between medical-technological possibilities and the ethical evaluation of them has become so great that traditional ethics has become powerless.[7] Slightly less strong is the oft-heard complaint that ethics always lags behind the latest developments and does not provide any guidance.

---

6. Rahner, *Schriften zur Theologie* VIII, 289. The meaning of the word "manipulation" is not necessarily negative. See Häring, *Manipulation*, 1ff. Häring shows that manipulation happens in all kinds of contexts (such as education, the media, advertising, economy, politics, and election polls). We must distinguish between manipulation of things and of the biological "nature," on the one hand, and the manipulation of the human person "in his inner sanctuary of freedom," on the other (44).

7. *Beter dan God*, broadcast March 15, 1987 (M. A. M. de Wachter).

Genetic Engineering (1)

However, it is not fair to accuse ethics of always lagging behind. In the first place, it is naturally the case that ethics must always *come after* the developments it must evaluate. When DNA had not yet been discovered, there could not be an ethical evaluation of DNA. And when, within a relatively short period of time, an avalanche of questions comes to us about new developments, it is no wonder that ethics needs time to reflect on the issues. It is not easy to survey everything that is or may come to be involved in genetic engineering. Your head spins when you delve thoroughly into this subject.

Second, there is the question of what people understand "ethics" to mean exactly. Many people believe that ethics should adapt to newly acquired knowledge. For example, when artificial insemination made it possible to have children with use of sperm from a donor outside a marriage relationship, many thought that such donor insemination should be ethically approved. When the topics of abortion and homosexuality became current, the same thing happened. Many ethicists who had previously raised objections dropped them, reluctantly or otherwise. It is no wonder that this kind of ethics easily gives the impression of lagging behind the latest developments. It is therefore understandable that people readily conclude that traditional ethics has become powerless because of the huge possibilities that have come within reach as a result of genetic engineering. The technology changes, and *therefore* traditional ethics, are no longer adequate.

The ethics that I am advocating in this series, "Ethical Reflections," is of a different kind. The ethicist must unquestionably remain up to date on all developments. Ethicists cannot act in the twenty-first century as if we were still living in the nineteenth century. Ethics has to be developed and corrected. But, *at the same time*, ethicists who let themselves be guided by God's revelation in Holy Scripture will be aware of *constant* factors that remain the same throughout the centuries and must be respected. If we have an eye for such constant factors, we may not say that ethics lags behind today. When ethics holds fast to what is constant, it simultaneously introduces order into a jumble of issues, also, when it involves developments in DNA technology. Even though ethics arrives on the scene after the issues it must evaluate (first DNA technology, then DNA ethics), something else is at least equally true: ethics precedes these developments, at least when it continues to maintain as constant what really is constant.

In what follows I will try to organize and discuss various issues involved in genetic engineering with the help of a number of continuing, constant principles.

THE "NATURE" OF WHAT HAS BEEN CREATED

In a previous chapter we concluded that God created plants, trees, and animals *according to their kinds* (Gen 1:11-12, 21, 24-25). This expression must mean that each of these creatures received from their Creator something that is unique. The special creature, "man," who was created "in God's image" is permitted to rule over other creatures, but then he must respect the unique nature of these other living creatures. They have been created in a plethora of forms.

Israel was told that it could not interbreed animals or plants (Lev 19:19; Deut 22:9, 11). Sexual relations between people and animals was also strongly forbidden (Exod 22:19; Lev 18:23; 20:15-16). Even though we no longer have to observe the Old Testament's injunctions against sowing one's field with two kinds of seed, or wearing a garment of cloth made of two kinds of material, it remains valid that the pluriformity of creation demands respect. People cannot do with it as they please. When they utilize the potentialities of creation, they must respect the individual nature of each creature.[8]

---

8. See Seldenrijk, *Genetische technieken*, 80ff.; and Bakker et al., *Landbouwbeleid*, 36ff., in which they make use of a number of elements from Schilder's definition of culture: to "develop [things] according to their own nature," to "make them serve the environment far and near," "in accordance with the cosmic contexts," and "pursuant to a self-imposed bond to God's revealed truth." For the definition itself, see Schilder, *Christ and Culture*, 40.

This does not mean that every form of engineering has to be rejected. As we saw, engineering occurred in the past as well. I do not think that there is a principal difference between what was achieved in the past, in the context of improving crops and growing new species, and what can be achieved in the future by way of genetic engineering. If crops can be improved via new methods and their much higher nutritional value can benefit humanity, we cannot object to modern genetic engineering.

When I claim that such engineering must be capable of benefiting humanity, for me the notion of "humanity" includes the understanding that the engineering of plants must benefit the poor and the hungry of this world as well. We must be offended by any engineering that ensures that only we in the wealthy Western world can obtain even more and cheaper food in an even more efficient manner. For it is certainly possible that biotechnology gets applied in the West but not in the Third World. The chasm between rich and poor will then only widen. In fairness we should note that, sadly, the widening of the chasm is already occurring in many areas. Yet we cannot absolutely reject genetic engineering solely on this basis.

The engineering of bacteria used for cleaning up the environment can also be a good thing. We do not have to reject such a possibility, as did the "Greens," the German environmental party. This party fears that the use of bacteria will only strengthen the tendency to continue to produce materials that are harmful to the environment and to our health. After all, people are apt to say, bacteria will clean up the mess we make. And surely the poisoning of the environment that results from our actions can be neutralized by engineered bacteria, can't it?[9]

Yet this kind of criticism contains an important warning. The environment has been polluted by people. To clean it up, *people* must change. It is not the case that they must change the *environment* for that purpose. When acid rain devastates coniferous forests, people should not make the trees genetically immune to

---

9. Commission of Inquiry of the German Bundestag, *Chancen und Risken*, 335.

such rain. Instead, they should stop their pollution. Bacteria can possibly assist in this, but the real disease cannot be fought with bacteria, but only by a different attitude on the part of humanity.

It becomes more difficult to justify genetic engineering when we are dealing with higher kinds of animals. It is a legitimate question whether the various plans being made for animal husbandry have not transgressed the above-mentioned boundary of "their nature." We are not speaking of therapeutic matters, which serve the welfare of animals. If genetic engineering for therapeutic reasons is good for people, it is good for animals as well. Think, for example, of improving resistance to such diseases as hoof-and-mouth disease in cattle.

However, genetic engineering of animals is focused not on therapy, but on our welfare. For example, some hope that transferring certain hereditary material to animals will enable them to digest vegetative material better. Animals lack the enzyme cellulase, which breaks down the hard cell walls of plants. That is why they have a very long intestinal tract that contains many bacteria that help to digest the vegetative material. If it were possible for animals to produce the digestibility enzyme cellulase themselves, that would entail unprecedented possibilities for feeding cattle.

This sounds wonderful, but are we then still respecting the "nature" of these creatures? Must we improve the digestion of animals? The fact is that the plans for this are intended for a more productive use of animals for our own needs. But if, for that purpose, one thing or another has to be altered in animals, I think this is alarming. Researchers are also trying to get cows to produce 16–20 percent more milk and to produce more meat by administering the hormone bovine somatotropin (BST). The cow's gene that is responsible for the natural production of this growth hormone is isolated and multiplied via recombinant DNA technology. When this hormone is then injected into the cow, it will lead to higher milk production. If scientists are also successful in improving the cow's metabolism, milk production can be increased even more.

The same applies to pigs, which grow more quickly and produce less fat with the similar hormone PST (porcine somatotropin).

Genetic Engineering (1)

But is that not an assault on such animals, which thereby become purely a product of our animal agriculture industry?[10] What kind of attention are we then devoting to the welfare of the animal itself if the BST indeed stimulates milk production, but also pushes the cows to their limit[11] so that they become more susceptible to fatigue and illness?[12]

Similar questions arise in the context of reproductive technology for cattle. It all started with artificial insemination. Recently, embryo transplantation became possible. A top-quality cow is treated with hormones to achieve simultaneous maturation of the ovaries. The ovaries are impregnated by means of artificial insemination. After approximately six to eight days, four to ten embryos are washed out of the cow's uterus and then transplanted into the uteri of inferior cows, which then function as surrogate mothers.

The offspring are of excellent quality, of course, because the hereditary material came from a top-quality cow and bull. But the normal reproductive process is bypassed. And that is the case even more so when one takes it a step further, as happens with *cloning* embryos. The cell nuclei are removed from a cow's embryo of sixteen cells. These cell nuclei are completely identical. They are transferred into ovaries from which the original nuclei have been removed. The result is that one embryo has become sixteen, so once the calves are born all have exactly the same characteristics,

---

10. These examples are from Van Dijke, "Manipuleren," 84, which contains criticism similar to mine.

11. This according to Meijer, in the unpublished paper "Biotechnologie bij het rund," a survey of opinions of members of the GMV (Gereformeerd Maatschappelijk Verbond [Reformed Social League]) in the milk farmers sector, Zwolle, 1988. All twenty-nine members were polled and all answered the following question positively: "Would you be pleased if the European Union decided to forbid the use of BST?" In fact, there is general hesitation about BST. "It is probably only the manufacturer of BST who really profits from this new product," according to *Biotechnologie, uitdaging of bedreiging?*, 19. BST can be compared to the human somatotropin (see the next chapter), but as bovine somatotropin, it cannot be used for people.

12. *Editor's note:* Since the publication of this work in the original, the use of BST was banned in the European Union in 2000.

such as the length of their tails, hair growth, milk production, meat production, etc.[13]

This raises a number of questions. If this development continues, the reproduction of cattle will likely occur exclusively by means of embryo cloning. But does not ethics then come to stand in opposition to economics? Understandably, cattle farmers want to have the best of the best. And they are forced to, for otherwise they may as well close down their operations. For there is already an excess of milk and meat, and the extent of livestock farming really needs to be reduced because the environment is being polluted with too much manure. But with cloning, livestock farming is *on the road to uniformity*. Only a few parent animals will ensure a massive production of genetically pure milk and meat products. The rest can be slaughtered.

This uniformity can lead to genetic erosion, an impoverishment of the gene pool. Many geneticists fear that genetic engineering will render the hereditary base too small. That applies just as much to crops in the context of arable farming. Genetic improvement must remain coupled with genetic variation. And do not our actions with regard to modern reproductive technology used in stockbreeding erode the individual variations given by God in creation?

In the second place, we must ask whether it is permissible to change the character of reproduction to such an extent that it becomes completely asexual. Someone could reply that we are talking only about animals. But who gives us the right to exclude the natural process of reproduction so radically, as is being threatened for domesticated farm animals? Years ago a professor condemned artificial insemination because we ought not to deny animals sexual gratification.[14] While his students probably chuckled about this, should we not take this criticism seriously? We certainly do not

---

13. Embryo "cloning" is distinguished from embryo "cleavage." The latter involves the division of an embryo into two halves, instead of allowing it to separate into single cells. See Van Leeuwen, in *Boerderij/Livestock Operation*, 30–31.

14. J. Waterink at the Free University, according to Manenschijn, *Geplunderde aarde*, 24.

Genetic Engineering (1)

have to condemn artificial insemination in every circumstance.[15] But when artificial insemination, embryo transplantation, and cloning completely replace normal reproduction in animals, it raises an important ethical question. Should we not reject these developments on the basis of the structure of God's creation?

Besides, what is then left of the good relationship the Bible demands of people toward "their" animals? Modern stockbreeders begin to look less like the righteous who have regard for the life of their animals (Prov 12:10) and more like dictators who do as they please to their subjects.[16]

PATENT RIGHTS FOR LIVING ORGANISMS?

Engineering of plants and microorganisms raises other questions as well. As we saw, an engineer sought a patent for the "cleanup" bacterium. The application was granted for that development—not of an inanimate thing like an internal combustion engine, a washing machine, or an electric razor, but of a *living* creature (no matter how insignificant a bacterium may be in the huge world of fauna). A gene was added to many different kinds of bacteria, not through natural mutation, but through human intervention. Essentially, the engineer sought to patent a form of life. Is that the direction we are headed: that patent rights are extended to vegetative, animal, and possibly human organisms, which become new kinds of property of researchers and manufacturers?!

15. See my work *The Ten Commandments*, 252–53. Van Dijk and Huygen, "Ethiek," 34–35, point out the following difference between artificial insemination and embryo transfer: while artificial insemination was to a large extent concerned with solving the problem of venereal infections, embryo transplantation can hardly be regarded as a solution to an existing veterinary or social problem. See also Meijer, "Biotechnologie," 21: "In nature it is typically the case that one male animal services the entire herd. This is often the strongest animal that commands the greatest respect. The cow is the limiting factor. She births only one calf per year. In reality you do the same thing with artificial insemination. The most optimal bull inseminates a large number of cows, but they still each birth only one calf per year. Embryo transplantation bypasses this law of nature."

16. Seldenrijk, *Genetische technieken*, 100.

Still, the U.S. Supreme Court did not grant the patent of the bacterium in question without giving the matter serious consideration. The Court realized that the matter involved potentially serious consequences, for it noted that Congress was free to rewrite the patent legislation in a more restricted way.[17]

Despite this circumspection, we should ask whether the patent did not go too far. Was not the fundamental difference between animate and inanimate material disregarded in this case? Is it possible that people can acquire ownership rights over genetic material and therefore over life, since that (bacteriological) life did not *originate* naturally, but because people *formed* it as its architects and builders? Is this formation really not a distortion of life? Indeed, the inheritance of living organisms does not belong to individuals or businesses; it is God's creation and, as such, has been given to people as a gift to be preserved and administered by them for the welfare of the entire human race.[18]

THE DANGER OF EXPERIMENTATION

Many people are understandably concerned about possible negative outcomes from these new, genetically engineered life forms. Suppose that a dangerous plant species comes into being, or that an engineered bacterium escapes the laboratory, replicates very rapidly, and then causes enormous destruction or illness. What kinds of dangerous things are people actually creating when they fiddle with hereditary material?

It is to the credit of biologists that researchers among them called for a moratorium—a temporary halt—on certain kinds of DNA research. That happened in America in 1974. Shortly before, a large group of scientists had gathered for the first Asilomar

---

17. Humber and Almeder, eds., *Biomedical Ethics*, 168–69. The American patent bureau declared that a (still imaginary) patent application for a new kind of human being will be rejected. See Seldenrijk, *Genetische technieken*, 95; and also Antébi and Fishlock, *Biotechnology*, 187ff.

18. Eibach, *Gentechnik*, 198ff. Unlike the United States, the European Union countries do not recognize patent rights to living plants and animals.

Genetic Engineering (1)

Conference, in California. The possible dangers of genetic engineering were broadly discussed at the conference. However, during the second conference (February 1975) the moratorium was lifted.

We therefore cannot say that biologists and other scientists are simply proceeding with their plans for genetic engineering without being very particular about safety measures, despite what others may think. That is an unfair and generalized judgment. Fortunately, some scientists follow the work of others very critically and also exercise great prudence in their own work.

The initial concern about possible major accidents appears to have been unfounded. Celebrities like Paul Berg and Stanley N. Cohen, who originally called for a moratorium, retreated from that position. Lab research can be conducted in great safety. For example, it is possible to "cripple" bacteria used in a laboratory. Such bacteria cannot live independently outside the laboratory, nor can they multiply in a person or an animal.[19]

Nevertheless, great care remains the watchword, for in this field, what seem to be small interventions can have consequences that are incalculable at the outset, but which later result in a disruption of the ecological system. We have already seen in our discussion of the environmental crisis that well-intentioned actions can ultimately have serious consequences.

But that is not, of itself, a reason to reject all genetic engineering. Risks attend every subsequent step that people take on the road to their discoveries. We cannot reject genetic engineering because it can be dangerous. That would be too easy a resolution of this issue. That is why I also did not reject nuclear armaments and nuclear power plants, even though both are accompanied by specific (and sometimes great) risks.

It is much more important to ask what the *goal* of genetic engineering is. And anyone who takes into account the nature and welfare of plants and animals to achieve that goal will undoubtedly exercise the greatest caution. We can learn from the past that we may not do to the environment whatever we wish. Everything is

---

19. Hermans, *DNA-onderzoek*, 36–37, 55–56, 62–63.

interconnected, and the consequences are great when the ecological balance is broken.

ONCE AGAIN: PEOPLE AS STEWARDS

Perhaps nowhere is the significance of the characterization of people as *stewards* as clear as with this issue. That term does not imply that people must leave the world as it is. We have been placed in the world to work it and to preserve it (Gen 2:15).[20]

The phrase "to work it" points to development and change. That scientists continue with DNA research is part of this development. It would be wrong to discontinue this research solely because it could yield frightening results. We did not stop when the printing press, the steam engine, and the airplane were invented, even though, in addition to having many advantages, these were accompanied by much misery as well.

In connection with this issue, when people remark that the truly legitimate reason for engaging in sound research is curiosity, we can agree.[21] It must be fascinating to learn all about the world of the living cell. We cannot stop science, nor may we discredit it with the assertion that science is solely concerned with profit, patents, and products. Even if not a single industry should pay the costs of the research, scientists would still discover the mysteries of DNA, though probably a bit later than is now the case.

It is human nature to investigate and constantly search further. We must not regard science as the enemy. Whenever science improves, assists, and serves life, it is a worthwhile endeavor.[22] And it can be worthwhile also in the context of genetic research, as we have seen.

---

20. See chapter 2 above.

21. H. Galjaard, in *Beter dan God*, March 15, 1987, 18.

22. See Hübner, *Die neue Verantwortung*, 106–7. In his opinion, a new ethics is not necessary. What is necessary is a new attitude in which people, pursuant to Christian conviction, begin to live *together* with plants and animals to the glory of God (108).

Genetic Engineering (1)

Abuse in this area is certainly a great danger. But we should not allow ourselves to be so overawed by sin that it seems as if there is no longer a creation with a mandate. For if attention is focused *right away* on the sinful misuse, scientific opportunities and activities will no longer be judged on "their own merits."[23] Creation and the abuse of creation are two different matters. Thus, we may launch our consideration of genetic engineering in a positive manner.[24]

But as stewards, people are also called to *preserve* creation. They are handling material that belongs to Another who has appointed them as managers. For that reason, we may have to respect certain limits to our curiosity, as we shall see.[25] People are not this world's creators, but merely its stewards. They may not tamper with the structure of creation, and their curiosity needs to be a *targeted* curiosity. What do they see as their task? They may intervene, but even then their attention must remain directed toward preserving creation. They engineer in order to improve what exists.

Therefore, we can expand the term "preserve" to include "protect" and "heal."[26] "Preservation" keeps us within the bounds of nature as we have received it from God's hand. "Protection" and "healing" presuppose that in many respects creation exists as

---

23. Thus, correctly, Huijgen, "Creativiteit."

24. This is why it seems to me incorrect to begin in a negative manner, as Schuurman did when speaking of "technicism" and "technicisation," and only then to conclude in a positive manner about "justifiable gene technology." See his address, "Wijsgerig-ethische," 7ff.

25. For that reason I disagree with the continuation of Galjaard's verdict in *Beter dan God*: "As far as I am concerned, you may investigate everything. I would not want to stop it simply because we do not know what the consequences will be." For example, someone who requires human embryos for research, embryos that will be destroyed in the course of the research, is going beyond the limits of curiosity.

26. I am following Eibach here, who uses the three concepts in *Gentechnik*: Bewahrung, Schutz, and Heilung (preservation, protection, and healing; *Gentechnik*, 90). See also his *Experimentierfeld*, 133ff. Variations are possible, of course. Thus, in his inaugural address, *Crisis*, Schuurman describes the motif that must guide farming in terms of the triplet "harvest, maintain, and preserve" (28).

an endangered and damaged entity. The way things are running now is not what God intended. Humanity's sin has damaged the world. Creation was subjected to futility and all parts of creation are groaning (Rom 8:20–23). People may and can do something about that by protecting themselves and their environment. They can try to heal the wounds. They can guard against the extinction of species, and they must seek to rejuvenate the deteriorating environment. They can detect diseases and try to eliminate them, even though we can never banish suffering and death.

This definition of the task of people as stewards seems conservative, and it is, but in the best sense of the word. It keeps us from the delusion that we can create or recreate our own world. In the context of preservation, there is all kinds of room for scientific curiosity that is not satisfied (from a misguided sense of conservatism) with what has been discovered but always wants to go further. The deeper that researchers delve into the secrets of life, the more they can protect and heal humanity and the environment. Indeed, necessary scientific research will always have a distress-relieving character.[27]

This preservation implies that we must forgo the desire to mold nature into our image. A mouse, a sheep, and a goat must remain what they are by nature, without any manipulation of their genes. We should not genetically alter cows and pigs in order to extract more from them than good breeding methods already give us. To use a more frivolous example,[28] we must not supply apes with human genes so that they can pick our apples in our place. God gave livestock to us, but he did not create the apes to become our fruit pickers.

Animals must remain animals and human beings must remain human beings. Consequently, any hybrid between a human being and an animal must be rejected. Theoretically, some think it is possible to create a new embryo by merging a human embryo

---

27. A beautiful remark of Hübner's, in *Verantwortung*, 112. In German the wordplay is even clearer: "Notwendiges, weil Not wendendes Fortschreiten steht gegen Fortschritt als Selbstzweck."

28. See Seldenrijk, *Genetische technieken*, 30.

Genetic Engineering (1)

with the embryo of a gorilla or a chimpanzee. But we should reject all experiments designed to create such chimeras.[29]

Hybridizing species will probably never become possible, or at best only in the distant future. Chimeras, which consist of a lion at the front, a goat in the middle, and a dragon at the back, existed only in ancient mythology. Just like a trip to the moon is but a small step into the cosmos, so too genetic engineering today stands at the starting line. Tinkering with genes is much more complex than building with Legos, despite the fact that they are sometimes compared to each other.

Theoretically, of course, everything seems possible, but reality is more intractable. Genetic engineering is already difficult simply with plants. Plant chromosomes are so complex that the technology used to isolate their genes is not suitable for other forms of life. The barriers are so high here that success, even in the long term, is unlikely save in very exceptional situations.[30]

However, people are capable of much, because they were created as almost divine (Ps 8:6). And in their demonic form they can achieve feats that earn them the number 666 (Rev 13:18)—not quite the number seven, which, as the number of perfection, remains reserved for God.[31] That is why it is good for our ethical reflection to let our imagination loose and then, taking into consideration what *might* some day become reality (even if it is highly unlikely), make clear what is not allowed—not only later on, but also not now, even if we are only talking about taking first steps. People endowed with their immense capacities must remain modest as stewards who work, preserve, protect, and heal what belongs to their Lord and not to themselves.

Only within this framework does genetic engineering make sense. I have not rejected all forms of genetic engineering with

---

29. See Commissie van de Gezondheidsraad, *Advies inzake kunstmatige voortplanting*, 99–100.

30. Thus De Vries, "Genentechnologie."

31. In fact, this is only one of the exegeses of a difficult text. See Greijdanus, *De Openbaring*, who regards the number 666 as the fullness of the world in its highest development, complete cooperation, and utmost exertion.

Environmental Stewardship

regard to plants and microorganisms and even animals, as long as it does not harm, but rather enhances, the animal's well-being. The species boundary is not violated if the transgenic animal has received an extra gene from the same species. It is different when genetic material from a rat is transferred to a mouse, or if we believe that we must improve the digestion or eliminate the sexuality of domestic animals. How can that possibly fit within the framework of our task to preserve, protect, and heal? We could imagine that it will protect and heal a hungry world, but in reality it is aimed at enhancing our own welfare even further.

Every intervention must take place with appropriate care, but that does not necessitate a total rejection. If it is "merely" a step further in the improvement of plants, why should we refuse to endorse it? This is also true of possibilities that serve beautification or improvement, even though they do not belong among the necessities of life. I think, for example, of the development of new colors for petunias, carnations, roses, etc., and of the production of artificial chymosin, which can reduce the ripening time for cheese.[32]

Speaking more generally, we must not be guided in terms of everything that is possible, but in terms of care and love for a creation that has been entrusted to us as stewards. It can be said about farming, "In harvesting food, in connection with which nature is carefully preserved and cared for, fertility is carefully preserved and promoted, and the landscape is managed responsibly, farming serves to provide the necessities of life for all people. At that point, science and technology are not taking precedence, but function in service to farming."[33]

As a matter of fact, that is always the function of science, also when it involves the genetic engineering of human beings. That is what we will discuss in the next chapter.

---

32. Schilpzand, in *Boerderij*, 31. In the Netherlands, coagulant for cheese is prepared from the stomach of newborn calves, which have become scarce and expensive. Engineered bacteria can now produce the enzyme chymosin (rennin), which reduces the ripening time and the cost of cheese.

33. Schuurman, *Crisis*, 35.

5

Genetic Engineering (2)
Human Beings

ARRANGEMENT

IN THE PREVIOUS CHAPTER I devoted attention to genetic engineering in the world of plant and animals, although I also noted the relevance of that issue for human beings. Genetic engineering of human beings, however, calls for a separate discussion. We encounter special problems in this context, problems that are not as important in the context of plants and animals. To bring some order to the extensive material, I will discuss the following topics:

1. genetic diagnosis
2. genetic selection
3. genetic therapy
4. genetic eugenics

Environmental Stewardship

1. GENETIC DIAGNOSIS (GENE MAPPING)

Increasingly, the mysteries of the human genome[1] are being unveiled. More and more, the genes that cause specific diseases are becoming known. No one really doubts anymore that within ten years a kind of gene map (also called a "gene passport") will be available for each person. It will enable the precise determination of an individual's genetic makeup.

When we reach that point, we will have achieved a very significant accomplishment. I have already pointed out that there are some three billion rungs on the DNA double helix of the human cell. Several combined rungs of the ladder form a gene. Human beings have between 50,000 and 100,000 genes, which form the foundation of our hereditary characteristics. Understandably, mapping all of them will be a long and complex process.

But the research is progressing. *Time* magazine spoke about "the gene of the week." Every week new discoveries are made in the world of the human genome. Someone who says today that the gene responsible for a particular disease is not yet known will have to announce tomorrow that it has been discovered.[2] And there are already devices that can map hundreds of thousands of rungs each day.[3]

The molecular biologist and the computer work together. Very soon, it seems, all our genes will be registered. A simple blood sample, a computer printout, and we will know our genetic composition. Science has penetrated more deeply than ever before into the mysteries of the human body.

---

1. "Genome" refers to the aggregate of all the genes of a particular organism (in this case, of human beings).

2. Schellekens and Visser, *Manipulatie*, 136–37, discuss cystic fibrosis or pancreatic fibrosis, the most common inherited disease in our region, from which one out of 1,600 children suffer. It is not yet known which specific gene is responsible for this disease, they write. But their book was barely published when Schellekens reported on television, in the program *Beter dan God: Een jaar later*, that the genetic location for cystic fibrosis has been found. Patients with cystic fibrosis (also known as mucoviscidosis) have chronic bronchitis and problems with the pancreas and liver.

3. Seldenrijk, *Genetische technieken*, 52.

Genetic Engineering (2)

It would be wrong to question the significance of the gene map by saying that health and sickness do not depend entirely on our genes. Environment, nutrition, and lifestyle also play a major role, do they not? Surely the genotype does not correspond to the phenotype?[4] This is true, and so it is incorrect to say, "Tell me what your genes are and I will tell you who you are." For genes do not disclose everything, and there is interplay between genes and environment. One person suffers a particular illness but another does not, even though they may both be genetically predisposed to it. We are more than our genes. We should not subscribe to genetic determinism, as if our future were hidden in our genes and all we have to do is look at our gene map to see what will happen to us.

Nonetheless, the increased knowledge of our genetic composition is very important. We are what we are in part because of our genes. Their discovery does not disclose *everything* about our future, but *what it can tell us* is very far-reaching. The gene map will confront us with immense problems. Perhaps no single aspect of the whole complex of genetic engineering will cause as many problems as the *knowledge* that our gene map gives us.

Consider, for example, a patient who consults a medical doctor and is told that his genetic pattern suggests Huntington's disease, which will cause the nervous system to deteriorate slowly after age forty. Or the doctor may conclude that the patient is likely to develop cancer later in life. Or again, the doctor may conclude from the gene map that the patient will at some point develop senile dementia. Some people, in fact, think that we will soon be able to predict a person's lifespan from genes.

The Right to Remain Uninformed Is Relative

When we understand all this, we are apt to say, "I do not want to know all that. I am not going to live my life in terms of the stars or the lines in my hand, nor in terms of my genetic pattern." Fair

---

4. "Genotype" refers to the hereditary composition of an organism. "Phenotype" refers to all the physical, observable characteristics of an organism, as they are determined by genotype and the environment.

enough, because there is also a *right to remain uninformed*, as we say nowadays. We are not obligated to know all that can be known about ourselves. Many people will be happy to remain ignorant about their genes. The literature even calls this right to remain uninformed a starting point for forestalling the difficulties that are likely to arise in this context.[5]

Indeed, we do not need to know what might possibly happen to us thirty years from now. No one may demand that we allow ourselves to be genetically screened. Although we know our lives to be safe in God's hands, the indeterminacy and open-endedness of the future are also matters that a Christian values. I remind you of what Jesus said in the Sermon on the Mount: "Therefore do not be anxious about tomorrow, for tomorrow will be anxious for itself. Sufficient for the day is its own trouble" (Matt 6:34). Why then should we need to know today what could happen to us tomorrow?

Thus, this right to remain uninformed becomes very valuable. The government will have to pay attention to it in its legislation, now that gene mapping is becoming a reality. We must avoid compulsory genetic screening of the population. Hundreds of thousands of people would be trapped in a state of uncertainty or perhaps even driven to deep despair if they had to cope with the knowledge that they could develop a serious disease within a certain amount of time.

And yet, this does not present the full picture. Suppose that it becomes possible to prevent diseases that we might develop later in life. Take the case of a person who might be susceptible to Huntington's disease, but would *not* acquire it if genetic-therapeutic treatments became available to prevent it. If you seek treatment early—at age twenty, for example—you could be treated successfully. But if you do not take this opportunity, is it evidence of unconcern when you rely on your right to remain uninformed? Or should we conclude that your attitude in this situation is evidence of carelessness?

I believe that the latter is correct. Prevention is surely better than suffering from a very serious disease, isn't it? For that reason,

---

5. Thus, for example, Van den Daele, *Mensch nach Mass?*, 79ff.

genetic screening for this and other potential diseases is likely to become commonplace. I used Huntington's disease as an example, but we can also think of all forms of cancer. Genetic predispositions can be identified for these too. The gene for intestinal cancer has already been discovered.

Of course, it is very frightening to learn that you are genetically predisposed to intestinal cancer. But, on the other hand, if you have this knowledge, you can have regular medical checkups so that an operation is possible before the cancer progresses too far.[6] If you have the potential for intestinal cancer, should you not be grateful that you can take timely steps to treat it? The right to remain uninformed does exist, indeed, but it is not difficult to think of situations in which we have a moral duty to know something. For then we can prepare ourselves for the evil day.

The result of genetic screening should not make us anxious. Jesus' words that I quoted above apply also after a bad result. Jesus does not say that we are not *allowed* to know what we are *able* to know about tomorrow. He says that we must not be anxious about what we may have to face tomorrow, or in twenty years. Also, when we receive disappointing news, we must deal with it in a Christian manner. When you increase knowledge, you increase sorrow, said the Preacher (Eccl 1:18), and that is likely to be the case for many in this context. But that is no reason to reject gene mapping—certainly not for Christians, who believe that their lives are secure.

Nonetheless, the right to remain uninformed retains its precious value. Knowledge of our genes should be restricted to tests for diseases that can be treated prophylactically. It will be beneficial to have a trusted medical doctor who will not disclose more things about our genetic screening than we need to know. Doctors ought to disclose problems only in *those cases* in which treatment is possible. That would be a good compromise on this difficult issue, although it is also not without its problems.[7]

---

6. This example comes from H. Galjaard, in *Beter dan God: Een jaar later*.

7. The confidential position of medical doctors presents problems. What kinds of responsibility do doctors assume? Do doctors have to maintain the

Environmental Stewardship

We can count on the fact that DNA diagnosis will play a very important role as soon as it can provide reliable and easy prognoses about our health. We must prepare ourselves for that. Many will get a better understanding of the fragility and finiteness of their lives than is possible with current diagnostic techniques. We should not wish to avoid this, for we are stewards also of our own health. The diseases we suffer now, as well as those we may possibly acquire, even though only in the distant future, impose a responsibility on us that we may not shirk.

Medical Examinations, Insurance, and Data Banks

Perfecting genetic diagnosis is likely to cause more problems. An adult can invoke the right to remain uninformed. But what will happen with routine examinations, which can soon be expanded to include complete genetic monitoring?

That is already happening in the case of *babies* who are examined at the health clinic. They are already being screened for certain kinds of hereditary diseases, like PKU (phenylketonuria), with the well-known heel prick. Such a limited screening may well become a comprehensive one in the future. Not only diseases for which treatment is possible, but also untreatable ones will then be discovered, resulting in all kinds of misery for the child, as well as for its family, so that they are no longer able to enjoy, in an uninhibited manner, the years before the disease is manifested.

There is also the specter that *businesses* will require employees to be screened genetically. In themselves, medical examinations are unexceptional. Everyone understands that they are necessary, also,

---

doctor-patient relationship in such a way that they inform X as soon as illness Y, to which X is susceptible, is treatable, a susceptibility to which a doctor may know on the basis of X's gene map? And if a doctor fails to do so, can X then bring legal action against the doctor because X did not receive timely treatment? In fact, we have to realize too that an individual's decision not to be informed is complicated by the fact that we are connected to other family members. Their gene maps also have some, or perhaps much, information about our own. If some family members prefer to remain uninformed, others can tell them of potential or suspected problems!

for the protection of the person who applies for a job. Someone suffering from bronchitis should not work with cement, and someone suffering a hernia should not be hired by moving companies to do lifting. But it is not impossible that simple medical examinations that look for *existing* conditions may soon be expanded to include genetic screening that will uncover *future* diseases. It is no wonder that such examinations are attractive to businesses that consider only their economic advantage. The better the health prognosis of a job applicant, the better for the business. But you can guess the consequences. Thousands of people will be unemployed, not because they *are* sick, but because they might possibly *become* sick.

Serious objections are quite properly being raised against compulsory gene mapping for employees. Employees would thereby figuratively stick their heads in a genetic noose. Many people are already asking governments to place limits on the right to insist on medical examinations.[8] The examinations should be restricted to measuring the health of the patients at the time of the examinations and should not involve any genetic analysis that looks for possible future diseases.

When we assume that genetic screening will soon become technically very easy to conduct, legal restrictions could easily be circumvented by means of a simple blood sample. Employers and examining doctors must act honorably and avoid acting simply in the interests of the business, that is, doing a complete scan, not disclosing it, and if the results are unfavorable, simply not hiring the applicant.

The possibilities of gene mapping are likely to play a role in insurance applications too. We can readily insist that this should not happen. But suppose that, on the basis of your gene map, you know of the possibility of a serious disease within the foreseeable future and you then insure yourself for several million dollars. It

---

8. Thus the Commission of Inquiry of the German Bundestag, *Chancen und Risken*, 168ff. The right to "informationelle Selbstbestimmung" (informational self-determination) must remain protected. Complete genetic examinations must not be allowed. However, genetic medical examinations conducted for a *limited* purpose, and in the context of health care, are permissible, if properly regulated (170).

seems to me that this can be avoided only if the insurer is permitted to require a genetic test, especially when the insured amount is particularly high. Is that not similar to what is already happening in the Netherlands? Think of the HIV positive test that an insurer may require if someone wants insurance coverage above two million dollars.[9]

I have provided two examples of situations in which third parties have a definite interest in our genes. There are likely to be more interested parties. Think, for example, about *the government and the courts*. In the United States, data banks of genetic fingerprints[10] are being developed. The state of California, for instance, is going to register the genetic information of prisoners. The American weekly magazine *Time* concluded, without any reservations, that American scientists will discover the genetic basis for alcoholism within five years.[11] If a predisposition for addiction and crime can be found in an individual's genes, arguments in favor of data banks for these categories will certainly be advanced.

Yet we do not have to see dangers everywhere. In the past, imminent apocalyptic futures have been prophesied, but they did not materialize. The truth is much more complex and intractable than science fiction would have us believe. The route to the dictator who can manipulate everyone because he knows everything about everyone is a very long one. But we do need to keep our eyes peeled.

Databases that include all of our genetic information are technically possible with the use of computers. They will undoubtedly be attractive to many organizations, just like central registry offices that store personal information have been. Information contained in the gene map is worth its weight in gold for many people. But at the same time, we realize how dangerous it is when others can gain

9. Ibid., 174; *Beter dan God: Een jaar later*, broadcast on March 27, 1988. See also my *Aids*, 52–54.

10. Obtaining a DNA fingerprint (already being used to identify persons in rape cases) is actually a much simpler process than a gene map.

11. Information provided during the program *Beter dan God: Een jaar later*, March 27, 1988, 21.

access to such information. Thus, our privacy could be seriously compromised if our genetic information was placed in a database. Therefore, we should regard with the greatest suspicion every step along the route to centralizing genetic information.

2. GENETIC SELECTION

Why the suspicion? Because the technology will make it possible to select people, or rather, deselect them. As has been said, this will result in genetic discrimination.[12] This is not discrimination on the basis of the color of one's skin or race, but on the basis of one's "genetic predisposition," which leads to "social invalidation." If you have a bad gene map, you will have a tough time in society. You do not need to be sick; what matters is the prognosis that you may become sick.

Clearly we are disinclined to adopt this new form of discrimination today. We might think that it probably will not reach that stage. Unfortunately, another kind of discrimination is already taking place. And on the basis of gene mapping, such discrimination will be able to be done more effectively than has been possible until now. I am referring to the selection involving unborn lives.

We adults already have our place in society. Younger people may well have a more difficult life than used to be the case. But *actual* rejection on the basis of a genetic diagnosis is currently possible only for unborn lives. Prenatal research permits making predictions at increasingly earlier stages about the health of an unborn child. Echography (ultrasound; ultrasonography) and amniocentesis are already possible in the sixteenth week of pregnancy, and chorionic villus sampling (chorion biopsy) in the eighth week. Continuing improvements in the diagnostics will permit the discovery of ever increasing abnormalities in the earliest stages of embryo development.

The earlier that specific abnormalities are detected, the better a person can prepare for the beneficial treatment of the child.

---

12. J. Retel in ibid., 7.

Increasing medical knowledge and skill will improve therapeutic possibilities for prenatal intervention. The unborn child is already a patient, for whom some interventions are already possible.[13] But to intervene we must know exactly what the situation of the unborn child is. Such prenatal research is, therefore, of great benefit.

But there is another side to this as well. Sadly, in most cases, if there are clear or probable defects a decision is made to abort the unborn child.[14] Prenatal diagnostics do not then serve the child's welfare, but serve the practice of selection instead. What is suspected gets eliminated. This will become more efficient once it becomes possible to map the genes of the unborn child at an early stage.

Everyone would like to have healthy children. It is already common to abort a child if the child suffers from Down's syndrome or spina bifida. It is quite predictable that those who do not want a child with Down's syndrome will soon not want a child with a cleft palate either, once that condition can be diagnosed via gene mapping.

Our judgment about this has to be unambiguous; our actions in this matter must be consistent. Those who take the first step will also take the next. If you want to prevent yourself and your child much misery by proceeding with abortion, the same will happen when you want to prevent lesser misery, even if it is just a matter of a cleft palate.

Rejecting abortion will become increasingly difficult, in my opinion. For the more infallibly the medical specialist can determine what is wrong with the embryo, or what might be problematic in this human life forty years later, the more courageous you must be to allow such a child to be born. It takes no courage to reject data banks. But greater courage will be needed to welcome the unborn disabled child exactly for what the child is: a helpless

---

13. Seldenrijk, *Genetische technieken*, 60.

14. Galjaard et al., *Voorkomen*, 148. Today approximately fifteen thousand sets of parents approve a prenatal diagnostic examination. A few hundred parents receive a negative report, which usually leads to abortion, according to a news item in *NRC/Handelsblad*, November 29, 1988, 17.

person, but no less a *human being* who already exists among us and may count on a loving reception.

We do not choose disabled children. Being pro-life does not mean being pro-suffering, as though the suffering of disabled children as well as of their families is not a trial. But we must deal with this suffering, and not eliminate it. Being pro-life means being pro-love, because with love we must protect and care for those who are weak and defenseless.[15] The ancient command "You shall not murder" remains valid, even though the most modern diagnostics make abortion more alluring than ever before.

3. GENETIC THERAPY

It may seem that the topic of genetic engineering has only bleak sides. But that is not really true. There are already different *drugs* available thanks to recombinant DNA technology. I will give a number of examples:

Without recombinant DNA technology it would not be possible to provide hemophiliacs with the coagulation factor they need for their blood as efficiently as is now possible. Until recently, large quantities of donated blood were necessary to obtain this coagulation factor. That is no longer the case. Now an extra piece of DNA is inserted into a bacterium. This "new" bacterium is able to produce the human coagulation factor in large quantities.

Also, patients who suffer from acromegaly benefit from DNA technology. Acromegaly causes an excess of the growth hormone, which leads to enlarged nose, chin, ears, hands, and feet. This condition can now be controlled by the genetically adapted intestinal bacterium *Escherichia coli*. Approximately 0.264 gallons of bacterial culture can be produced in a couple of hours and will supply the same amount of the hormone as can be obtained from 500,000 sheep brains.

In addition to excessive growth, dwarfism also occurs. In the past it was difficult to obtain an adequate quantity of growth

---

15. I have taken the combination of the terms "pro-life," "pro-suffering," and "pro-love" from Van Bruggen, *Hulpverlening in gebroken situaties*, 6ff.

hormones to counteract the condition. This used to be taken from the human pituitary gland (hypophysis).[16] Now the remedy in question (somatotropin) can be manufactured through genetic engineering.

It is well-known that people with diabetes need to have insulin injections. This hormone is also now being manufactured with the aid of recombinant DNA technology. A small piece of the DNA that is responsible for the production of insulin is removed from the nucleus of a human cell. It is transferred into a bacterium, which then produces human insulin. A number of side effects that used to be caused by insulin (previously taken from the pancreases of pigs and cattle) no longer occur now.

We could also mention the product called interferon. Until 1980 large quantities of blood were necessary for its production. It is used to combat certain kinds of viral illnesses and some cancers. Vaccines are being manufactured with new methods too. The vaccine against Hepatitis-B, a form of jaundice, is very significant. Formerly, it had to be prepared from the blood of chronic carriers of the hepatitis virus. But that blood could also contain other causative agents of diseases like, for example, the pathogen that causes AIDS. Compared to earlier remedies, DNA products are much purer and more naturally produced.[17] Undoubtedly we will be surprised by more discoveries in this area, which will benefit increasing numbers of patients.

But there is more. Many people expect that somatic gene therapy will also become possible within a few years. Recombinant DNA technology holds promise for correcting genetic defects in cases of hereditary abnormalities. For the time being, this will affect only those defects that are related to a single gene. In gene therapy, a defective gene that causes a disease is replaced with a new gene that takes over the function of the defective one.

---

16. See my *Rondom de dood*, 134–35, for the ethical problem involved with this procedure.

17. See Seldenrijk, *Genetische technieken*, 27ff., from which this information is taken.

Genetic Engineering (2)

We may rejoice about such a development. Let us suppose that the thousands of innate and hereditary diseases have all been mapped, so that we know exactly where they are located on the DNA strand. It would then be possible, at the very beginning of life, to predict not only if the child has the extra chromosome that causes Down's syndrome, but also if the child has an inherited tendency for Duchenne muscular dystrophy, which will lead to paralysis in the child's early years, or whether the child could get Huntington's disease, and so on.

It may well become possible, with the aid of new technology, to intervene already in the prenatal stage to prevent all kinds of disabilities from developing. Defective genes will then be deactivated and healthy genes transplanted. Despite all we have to fear from these new developments, these are possibilities that we may accept gratefully. In the future, such developments could perhaps have an inhibiting effect on abortion. If the unborn child does not have to live with a lifelong disability, the so-called eugenic warrant for abortion disappears.[18]

Somatic gene therapy does not, therefore, raise ethical concerns. We can compare it quite properly to an organ transplant. Just as we consider the removal of an *existing* affliction by an organ transplant to be a blessing,[19] so too we can evaluate positively those micro-transplants that *prevent* human suffering.

But gene therapy is not without its ethical problems. Before it can be used, it must satisfy a number of conditions. A new piece of DNA must not be put in the wrong place in the DNA of the cell, as a result of which cancer cells could develop. The viruses that are used for the transfer of the DNA material must not cause damage to the person. We must also take into consideration the possibility

---

18. The improved technology already prevents a few abortions. Duchenne muscular dystrophy affects only male descendants. Because the gene that causes this condition has been discovered, DNA research can now also show whether a boy actually has the disease. Without the research he would probably be aborted ("safety first" is the argument). Now refined diagnosis can prevent it.

19. About which see my *Rondom de dood*, 117–38.

that precluding one particular disease can sometimes lead to the onset of another disease.[20]

Defective genes that previously were not passed on to a person's descendants may also become hereditary as a result of gene therapy. There are now diseases from which children die, but they will not have descendants with the same defective gene. But if they are cured through somatic gene therapy, then *they* do not have to worry about the disease anymore, but their *children* might pass it on.[21]

Furthermore, difficulties are apt to arise regarding which patients are to receive gene therapy. There are not yet any laboratory models for all kinds of diseases, and so there is no data about the feasibility of gene therapy. Consequently, human beings themselves will have to serve as "laboratory models."[22]

We should not condemn this in every instance. We can imagine that someone who is incurably sick and in the last stages of life may want to try a final, but not yet adequately tested experimental therapy. Who knows, perhaps gene therapy may be the answer! Of course it is still a question whether the reputation of the medical doctor eager for the success of the gene therapy, or the interest of the patient, carries greater weight.[23]

---

20. Sickle-cell anemia can serve as an illustration. This is an inherited abnormality in the red blood cells. The disease is widespread in Africa and it is not accidental that it is found there, for people who have sickle-cell anemia cannot contract malaria. Malaria is, of course, more serious than sickle-cell anemia. When you prevent one illness it can lead, as in this case, to another. Cell therapy, therefore, can lead to consequences that are unpredictable. Thus H. Schellekens in *Beter dan God*.

21. An example of this is ADA (adenosine deaminase) deficiency. Children with a hereditary shortage of the ADA enzyme are doomed to die. It is a disease that paralyzes the immune system. Because affected children typically die within a year, they will also not have descendants with the defective gene. Society will demand sterilization of patients who are cured in return for the undoubtedly expensive gene therapy. See Schellekens and Visser, *Manipulatie*, 144ff.

22. Schellekens and Visser, *Manipulatie*, 144.

23. Eibach, *Gentechnik*, 167. He asks whether fatally ill patients are really able to make decisions freely, and further, whether we ought not to undertake therapeutic attempts for such people unless there is a realistic chance of success.

Genetic Engineering (2)

Still, we cannot condemn somatic gene therapy on the basis of such difficulties. For similar difficulties occur even now. It has not always been possible to predict the results of drugs that have been allowed on the market. It is also true that for a long time already modern health care has allowed people to live longer, when previously they would have died years sooner without descendants with the same genetic defects. We are keeping alive not only sick people, but diseases as well. But that is no reason to allow disabled people to die.

As far as the risks of somatic genetic intervention are concerned, we always have to weigh risks. I repeat what I wrote about risks in the context of organ transplants: the risk of death may be great, but the certainty of death without a transplant is one hundred percent.[24] That applies also to gene therapy.

Germline Engineering

If applied successfully, the genetic therapy discussed to this point will cure people of inherited diseases. But the cure is only for themselves and not their descendants, as we saw. The correct gene is inserted into the somatic or body cells of an individual, with the goal of eliminating the consequences of an inherited disease in the patient. But the descendants do not yet benefit from this new gene.

It would therefore be better if genetic therapy could eradicate hereditary diseases completely, so that a patient's descendants would also be free of them. That is possible only if gene therapy is applied to the reproductive cells (the germline [sperm and ova, or gametes] cells).[25] For then the correction would apply also to cells that transfer genetic information to the next generation. Tackling

---

24. See my *Rondom de dood*, 136.
25. Such interventions can be imagined in the parental cells that produce reproductive cells, the reproductive cells themselves, or the cells of a pre-embryo (the stage in which the cells are still "pluripotent"). See *Advies inzake kunstmatige voortplanting*, 157; and Commission of Inquiry of the German Bundestag, *Chancen und Risken*, 184. Pluripotent cells are cells in their undifferentiated first stage.

the defect, then, literally begins "*ab ovo*," in the fertilized ovum, so that the correction can be carried forward in the reproductive cells. Techniques for this so-called *germline engineering* have already been applied in laboratory experiments on animals. There is every reason to suppose that is applicable to human beings as well.

What should we think about this form of gene therapy? Is it an extension of somatic gene therapy? Or is there a principal difference?

The consequences of germline engineering undoubtedly will be much more significant than those of somatic gene therapy. For in the former, you make decisions for future generations as well. For that reason alone, many people condemn this form of therapy. They agree that somatic gene therapy is permissible because it restores existing persons. But germline engineering is impermissible because it affects future generations. Conducting this therapy would mean that today's human beings are actually creating those of tomorrow. It has even been said that this would mean the definitive dominion of the current generation over future generations.[26]

I believe that we should not draw such an absolute distinction between somatic therapy and germline engineering. If parents are permitted to make decisions about the therapy of their child, even in the child's prenatal condition, why should they not be allowed to do so for their grandchildren? Suppose that everyone believes a hereditary disease may be eradicated in subsequent generations, with full accuracy and without negative effects, by germline engineering. What principled objections could we have to such a procedure? Are there not other diseases that we want to eradicate, like smallpox, polio, and malaria?

The argument that individual genes are given to us by God, and that we have to let our descendants be born as God wills,[27]

---

26. Commission of Inquiry of the German Bundestag, *Chancen und Risken*, 188.

27. Others speak about "nature" instead of God. The Commission of Inquiry of the German Bundestag does that and rejects gene therapy with the argument, among others, that human beings are not the design and the planned experiment of their parents, but the product of chance. This should not change, for it ensures the mutual independence of human beings and their

Genetic Engineering (2)

fails to understand the difference between the good creation of God and disease that results from the fall. We are allowed to exterminate diseases. Why then should the present generation not have a responsibility in that regard for future generations (plural)? Do we not assume a similar responsibility when we are talking about the management of the environment? The *individual* genes are genes of *humanity*, and decisions of previous generations often have consequences for generations that follow. For not every intervention in reproductive cells has to entail a change in the physical and psychical *integrity* of persons and their descendants. It depends on the purpose of germline engineering.[28]

More important is the argument that such an intervention will have consequences that are difficult to predict. If somatic gene therapy is unsuccessful, there is only one victim. But what shall we say of germline engineering, the results of which can extend across generations? You can determine whether such an intervention is justified only after a few generations, but then it could be too late. Who dares to take responsibility for a change that could have unforeseen consequences far into the future?

Tests that have been done on laboratory animals give pause for thought, for they have produced many failures and few successes. For example, a test with fertilized mouse embryos, which used three hundred animals, resulted in only eleven viable mice, and of them only six possessed the implanted gene.[29] Experiments in this area have only a slender chance of success. Who could then

---

individual self-worth (187).

28. In fact, that applies to somatic gene therapy as well. It can also go awry if we thereby alter the "integrity of the person." With similar arguments, Eibach, *Gentechnik*, 166ff., also correctly points to possible similarities between somatic therapy and germline engineering. The idea that somatic gene therapy is *merely* a form of transplanting does not automatically justify this form of therapy, for with a "normal" transplant we also have to consider what is being transplanted. Does it include reproductive organs, brains, or ovaries, assuming this is (or becomes) technically possible? See ibid., 168, and my *Rondom de dood*, 127.

29. Commission of Inquiry of the German Bundestag, *Chancen und Risken*, 188.

defend germline engineering when the subjects are not mice but people?

That brings me to the decisive point, namely, that which makes it impossible for us to approve of germline engineering. Let us assume for the moment that tests on mouse embryos will always lead to perfect results; this will still lead inevitably to the next step, namely, tests on human embryos. Germline engineering is simply inseparable from experimenting with embryos. But is it ever permissible to use human embryos as guinea pigs?

Those who object to abortion on the basis of God's command "You shall not murder" will also reject experimentation with embryos.[30] For these embryos that are relegated to serving as material for laboratory tests die as a result and are then destroyed. Use of embryos implies that they may be misused. These kinds of experiments do not benefit the embryos—their good development and their successful birth.

But when the embryo is viewed as a guinea pig, it no longer has a use in and of itself; it has only become a means for science to make advances in its research. No matter how noble the goal of such research may be, the means of reaching it are too costly.

Experimenting with Pre-Embryos

Sadly, many people think that we may experiment with embryos. From a research point of view this is very understandable. Without embryonic experiments, researchers will not be able to take optimal advantage of the scientific and technological possibilities that enable genetic engineering. Indeed, the possibilities may be precluded.[31]

---

30. This concerns experimental scientific research "that does not, or not exclusively, serve the embryo to be examined; one can speak of a total or partial instrumental use of the embryo." Thus the memorandum "Artificial insemination," by Korthals et al., 38. This experimentation is properly distinguished from actions in which the embryo itself is the object, and from mere observations of the embryo in the context of education (ibid., 37).

31. See Seller and Philipp, "Research on Human Embryos," 22ff. They list twelve topics for research, most of which will result in the death of the embryos.

Genetic Engineering (2)

It has been proposed that restricted use of embryos is possible. There are often extra embryos from *in vitro* fertilization[32] that are not implanted into the womb. Why may researchers not use those for experiments? That is surely something different than *creating* embryos by means of *in vitro* fertilization, isn't it? Besides, experimentation can be restricted to, for example, embryos younger than fourteen days.[33]

Even the terminology suggests that this concerns only a rather harmless matter. It is said that experiments are conducted on *pre*-embryos, that is, embryos between one and fourteen days old, when it is not yet certain what will develop from them. Is it not the case that approximately 50 percent of embryos are naturally and spontaneously lost during this stage? And is it not rather an exaggeration to speak of an "individual" or a "person" at this stage?

However, the boundary drawn here between the pre-embryo and the embryo is arbitrary. Of course no one believes that the pre-embryo is not yet obviously a *human* life. For the desire to have access to pre-embryos is based precisely on the certainty that *they are human* and not some other kind of life.

The story is told about Robert Edwards, who collaborated with Patrick Steptoe on the first birth resulting from *in vitro* fertilization, that he took the baby, Louise Brown, in his arms and then said spontaneously that he had last seen her in the eight-cell stage![34] With those words, he expressed exactly what no scientist will dispute: from conception to birth we speak of a continuing new human life. Indeed, Steptoe verbalized what J. Lejeune formulated this way: "A human being is born very tiny."[35]

---

32. About *in vitro* fertilization, see my discussion with W. H. Velema (who rejects *in vitro* fertilization) in Witkam et al., *In-vitro-fertilisatie*, 133ff. This matter, in addition to the whole complex of modern forms of procreation, is discussed in my *Seksualiteit en huwelijk*. See also my work *The Ten Commandments*, 249–60.

33. E.g., in the *Advies* of the Gezondheidsraad, 199ff. See also Korthals et al., "Artificial Insemination," 39, which proposes to forbid experimentation with embryos that have gone past the fourteen-day developmental stage.

34. See Jochemsen et al., *Menselijke embryo*, 21.

35. See my *Abortus*, 11.

Environmental Stewardship

It is nonetheless a legitimate question whether we can already speak of a new *individual* life from the moment of conception. For in the pre-embryonic stage it is possible that two individuals will come into existence from one embryo. Think, for example, of identical twins. It is also possible for two embryos to fuse and become one. But even so, why should the pre-embryo have less right to protection if, instead of one, two individuals may arise from it? And how can the exceptional case of the fusion of embryonic nuclei compel us to conclude that the pre-embryo is less worthy of protection? For in all cases *human* life is involved.

We can argue for a long time about concepts like "individual" and "person," but not about the fact that a person who destroys pre-embryos or embryos is destroying human life that could have grown to maturity. We do not yet refer to an acorn as an oak, or to an egg as a chicken. Therefore it is understandable that we do not refer to the pre-embryo as an individual or a person. But there can be no doubt that the fertilized ovum already possesses the full potentiality to develop into a human person, whose humanity begins in this pre-embryonic stage.[36] For that reason we must maintain a complete hands-off position.[37]

We can use and dispose of acorns from oaks and eggs from chickens, but not of human embryos. This point of view is based on a religious conviction, namely, that human beings belong to an absolutely unique class. God made them in his image and,

---

36. So also Kluxen, "Fortpflanzungstechnologien," 7. The *Advies* of the Gezondheidsraad correctly notes, "The commission's starting point is that every form of human life, no matter how early its stage, has an intrinsic value, because it contains unique and unrepeatable information for a (future) person and is for that reason entitled to protection" (74). But almost immediately it retreats from its strong position by saying, "If very important interests of many people are involved," then one "can consider whether by way of exception human pre-embryos may be used instrumentally" (75).

37. Therefore we should also not employ the argument that embryos may be used experimentally for the welfare of other embryos. Thus, Jones in *Manufacturing Humans*, 234ff. The writer, who gives instructive information from a Christian perspective about new reproductive methods, errs on a decisive point here.

Genetic Engineering (2)

accordingly, human beings must be treated as beings who are absolutely unique.

It has been stated, correctly, that assigning the labels "person" and "individual" to an embryo involves a confession about human beings.[38] Such terms are perhaps inadequate and not suited for the context of the pre-embryonic stage. But they are being used in order to demand protection for human beings in the very first stages of life.

If we do not insist on that, then an embryo's right to protection could be measured on a sliding scale: because the likelihood that a developing embryo will grow into a human being gradually increases, it must be treated with greater respect in accordance with its stage of development.[39] Apparently 50 percent of pre-embryos are lost in the natural process of pregnancy. And *therefore*, according to this argument, the pre-embryo is entitled to less protection than the embryo in later stages, when the percentage of loss decreases significantly and the likelihood of birth increases.

For those who regard human life as a special creation of God, regardless of the phase in which it exists, such an argument sounds just as strange as if someone should argue: because the likelihood that frontline soldiers will die sooner than reserve soldiers, many believe that the lives of reserve soldiers should be accorded greater respect than the lives of frontline soldiers.

Undoubtedly, life will be appreciated differently in its different phases. Our feelings for a newborn child are stronger than for a human embryo in its eight-cell stage. We also do not need to argue that every human life is of the same value. In addition to fundamental equality, there is also fundamental inequality among

---

38. See, e.g., Jochemsen et al., *Menselijke embryo*, 14, about the "transcendent mystery" of the human being, which cannot be supported or denied by science. Further, Long, "Infanticide," 79ff. It is striking what Long says about the relative meaninglessness of a dispute about "personhood" between people who have a totally different view of life: it is as if they are trying to drown a duck by aiming at it with a water hose (81).

39. This formulation is an almost exact rendering of what appears in the *Advies* of the Gezondheidsraad, 46, as the "opinion of many."

human beings.[40] Value and the right to protection are two different things. Two lives that are not of the same value are nonetheless both entitled to protection. We must accord equal rights of protection to *all* human life.

Theologians Are also Paving the Way

It is regrettable that even theologians are also paving the way for experimenting with pre-embryos. They do that sometimes by using the old philosophical and theological distinction between the as-yet-unformed (inanimate) and formed (animate) fetus.[41] During the first forty days after conception, you cannot not yet speak of an animate life, according to many theologians of the ancient and medieval church. It is obvious that those who held to this idea did not consider abortion during that period of time to be murder. However, they did consider abortion in that first phase to be a reprehensible act.[42] Still, they thought that those who committed abortion were more deserving of punishment if the fetus was older and "animate."

It is strange that theologians today use the ancient idea of an offender being *deserving of punishment* to advance their contention about an embryo's increasing *scale of right to protection*. What the church formerly called an offense, also during the first forty days, is now being approved by theologians to support current experimentation on pre-embryos. The consequence is that experimental research is being protected but the pre-embryo is not.

My position leads to the opposite conclusion. We may not experiment with (pre-)embryos and therefore we must condemn

---

40. I wrote about this more extensively in *Vrede in de maatschappij*, 22–24.

41. For example, Mahoney, *Bioethics and Belief*, 57ff., 96ff.; Dunstan, "Human Embryo," 39ff. See also Schroten, *In statu nascendi*.

42. See my *Abortus*, 41–46. Schroten does in fact formulate this belief correctly: "The seriousness of the offense depended . . . on the question whether the aborted fetus was 'formed' or 'unformed'" (*In statu nascendi*, 8). Exactly! It was always an offense, but not every offense was regarded as equally serious.

Genetic Engineering (2)

research of that kind. I realize that this rejection makes much important research impossible, for experiments with embryos are essential for the realization of germline engineering. But something else is more important for those who are totally serious about the respect we owe to human life.

In the previous chapter I noted that our curiosity is a valid reason to conduct research. But scientific curiosity must also respect boundaries. It cannot demand that it be able to treat corpses, as well as embryos, which are washed down the drain after experimental research, as mere objects of research.

4. GENETIC EUGENICS

Our rejection of every experiment with human embryos means that an important part of research is impossible. But that also has advantages. If everyone refrained from experimenting with embryos, the door to what many fear—*genetic eugenics*—would also remain closed.

There is a difference between negative and positive eugenics. *Negative* eugenics seeks to prevent bad genetic material from being born, continuing to exist, and being passed on to the next generation. I have already said what needs to be said about this in my discussion about abortion after hereditary defects are discovered in an unborn child.

While negative eugenics is concerned with the prevention of genetic defects, *positive* eugenics is directed at the improvement of the human species. In the context of our topic, this means the improvement of hereditary material in human beings by way of genetic engineering.[43]

---

43. "Eugenics" can also be taken in a broader sense as "the study—subject to societal control—of methods that may improve or affect the heritable (physical and psychological) qualities of future generations" (*Grote Winkler Prins Encyclopedie*, 380). Eugenics is already much older than genetic engineering. See Schellekens and Visser, *Manipulatie*, 15ff., about the origin and rise of eugenics. For eugenics in National Socialism, see ibid., 65ff.

Environmental Stewardship

To achieve that goal, germline engineering will be necessary, as will be apparent without extended discussion.[44] Those who reject germline therapy as ethically impermissible will therefore also have to object to positive eugenics.

But that is not the only reason why we must reject eugenics. Eugenics goes a step further than germline engineering. Eugenics is not concerned about healing human beings, as is gene therapy, but it seeks to refine them. To put it another way: those who promote eugenics do not want to make human beings better; they want to make better human beings.

It must be said that germline engineering, with its ongoing consequences for descendants, can easily turn into the planned refining of human beings. Where does disease end and health begin? Nowadays "sickness" encompasses a lot more than it used to. The World Health Organization regards health as "a state of complete physical, psychological, and social well-being, and not merely as the freedom from illness and infirmity."[45] If you take that seriously, you must come to the conclusion that really no one is healthy anymore. Who can say that they are physically, psychologically, and socially in a state of *complete* "well-being"?

In fact, then every limitation on our happiness already constitutes sickness and we ought to applaud everything that can reduce such limitations. Why then should we not make use of the enormous possibilities that recombinant DNA technology gives us? We have passed the threshold and are now able to intervene directly in our genetic pattern. Should we then not take the opportunity to become better human beings by means of the improvement and rearrangement of our genes? Why may we intervene for smallpox, polio, and sickle-cell anemia, and then not strive for higher intelligence and beauty, greater physical strength, greater ability to handle stress, and more altruism? If the goal is to improve the human condition, why draw a distinction between healing and

---

44. See, e.g., Van den Daele, *Mensch nach Mass*, 195.

45. See Eibach, "Gesundheit und Krankheit," 162ff., for a discussion of this definition of the World Health Organization.

Genetic Engineering (2)

refining?⁴⁶ And why not also try to multiply refined specimens of the human species via cloning?

We shall not pursue the question of whether it will ever become technically possible for eugenics to achieve what I just listed by means of genetic engineering. The complexity of the human genome is enormous. The elimination of defects that are "merely" caused by one wrong gene already presents immense challenges. That applies a thousand-fold for complicated processes like intelligence and psychological behavior. But I am sidestepping these problems because I am concerned about a principle.

Let me begin with the most far-reaching issue, the *cloning* of people. Cloning (there are other forms) involves the following process: the nucleus of an unfertilized ovum is removed and replaced with the nucleus of a specialized somatic cell of another living (or even dead!) person. The human being resulting from the altered ovum will then be completely identical to the person from whom the new nucleus originated. Cloning would enable, therefore, the creation of an entire mass of people having a completely identical genetic makeup.

The following fantasy is deliberate: if the person in question has the brains of an Einstein, the musicality of a Beethoven, and the limbs of a sports champion, we could produce supermen by means of continual cloning. But what may cause one person to rejoice will cause another to shudder.⁴⁷ Fortunately, most people still shudder at the idea of creating legions of "Übermenschen" (supermen).

It is obvious that we oppose all cloning of human life. We are what we are *individually*, with our God-given or God-withheld opportunities. Individual, infinitely different people must not be replaced by cookie-cutter people, no matter with what beautiful eyes, brilliant minds, and athletic limbs they might be supplied.

---

46. Singer and Wells, *Reproduction Revolution*, 182ff.

47. For this topic see, e.g., ibid., 150ff. Cloning is a completely asexual process and occurs completely outside marriage. This aspect is discussed in the volume in this series, "Ethical Reflections," entitled *Seksualiteit en huwelijk*.

We must not develop our own image of a human being in order to leave its stamp embedded in the genes of our descendants.

Many government organizations realize that watchfulness is required and that laws have to be modified to counteract disastrous developments. There appears to be agreement among a broad spectrum of people, both nationally and internationally, that the cloning of people, and the creation of chimeras, which could come into being through the combination of human and animal sperm and egg cells, are impermissible.[48] The rejection of such ideas is not always accompanied by an ethical justification.[49] But most people realize intuitively that positive eugenics puts human society on a dangerous path.

In the previous chapters we saw that God appointed human beings as stewards over creation. As part of that stewardship they were also entrusted with the management of themselves, including their genes. We also saw what this management involves: work, preservation, protection, and healing. We must protect what is threatened. We may try to heal what is defective, if necessary with the help of somatic gene therapy. The human genome is not taboo, as if it may not be the subject of research and, if necessary, be altered *therapeutically*.[50]

But we must clearly distinguish therapy from refining. Refining can be a good thing when it involves a plant or an animal, although there too we must impose limits, as I argued in the previous chapter. But as soon as we embark on the path of refining *human beings*—by making them more intelligent, stronger, endowed with longer life, more beautiful, and more socially gregarious—we are no longer engaged in "preserving" what has been entrusted to our stewardship. Rather, we cross the boundaries that have been

---

48. See Korthals, et al., "Artificial Insemination," 38-39.

49. That is always apparent in the reports of committees that have broad (pluriform) representation. A framework explaining *why* this or that is ethically impermissible (cloning, creation of chimeras, experimenting with embryos, etc.) is totally absent. See, for example, the *Advies* of the Gezondheidsraad, 99ff., 155ff., 199ff. The same lack is evident in the well-known Warnock Report, found in Warnock, *Question of Life*.

50. Eibach, *Gentechnik*, 179.

Genetic Engineering (2)

placed upon our stewardship. Then we are creating human beings after our own image, instead of accepting ourselves and others as we are, that is, as image of God and gift of God.

Accepting ourselves as we were created by God means that we accept the boundaries belonging to our *creaturehood*. We are human beings and no more than that. Every attempt to do better than God is hubris. We must also accept the boundaries that pertain to our *fallen* creaturehood. By our fall into sin we have brought suffering and death upon ourselves (Gen 2:17; Rom 5:12–14; 7:20).

We may combat suffering and death insofar as that is possible within the framework of *healing*. But any attempt to eliminate suffering and death would indicate overconfidence. Not all suffering is sickness, and even death is more than a genetically determined event that we should, by genetic engineering, defer and, if possible, eliminate altogether.

It is true that neither death nor the diseases that plague human beings are part of God's good creation. But they do belong to *our* fallen life that, in accordance with God's will, must come to an end. We cannot fight for bodily victory over death; that is a gift of Jesus Christ that he will confer when he returns.[51]

That must determine our attitude with respect to every attempt to eliminate death. The average life span has increased with better hygiene, better nourishment, and more effective medical treatments.[52] In that respect we show ourselves to be good stewards when we are busy protecting and healing our own lives. But we are not stewards anymore if we are systematically trying to extend our lives and to shake off our suffering and death.

---

51. See my *Rondom de dood*, 55–56, which criticizes H. Gödan and O. Jager, who want to mobilize science and technology against death. I shall leave aside the practical question of what the effect on society would be if we were to use eugenics to achieve longer life.

52. All of these do have "eugenic" *consequences* in that people live longer, are sick less often, suffer less pain, etc. However, positive eugenics does not work according to a medical model. Its program is an increase in intelligence, physical strength, beauty, etc. Even Singer and Wells admit that (*Reproduction Revolution*, 183).

CONCLUSION

When we try to survey the entire field of genetic engineering, the complexity of the subject can make us dizzy. The developments happen so fast that anyone who writes about it today must soon revise what they have written. Does this mean that we are ethically at our wits' end? Does ethics lag behind or, worse, is it so hopelessly out of date that it is therefore unable to respond to new developments? Should the ethical rules be changed and adapted to the new situation?

Without denying the novelty and complexity of genetic engineering, I believe that our Christian ethical reflection does not lag behind. Especially on decisive matters, it can employ *old* answers that have not lost their value in the face of turbulent new developments. In fact, I believe that the old answers permit a relatively straightforward ethical survey of this complex material, if we continue to adhere to the ancient concept of human beings as stewards of God.

God has entrusted the management of creation to human beings. They may *work* it and may continually dig deeper into the mysteries of creation. We can admit that every new solution to mysteries also brings forth new threats, without drawing the conclusion that science must stop its research and experimentation.[53]

As stewards, human beings may also be busy *healing*. Developments in the field of somatic gene therapy can alleviate the sorrow that often accompanies genetic diagnosis.

Furthermore, it is the task of human beings to *protect* weak human life that is entrusted to them. This limits their research

---

53. Accordingly, in my assessment of genetic engineering, I did not head straight toward the threat of a world with complete perfection, populated by tailor-made human beings engineered by super-brains. This would be an easy sort of prophecy, made apart from a good analysis of the subject. Only after we first admit that genetic research and genetic engineering are legitimate phases in the progression of science can we then come with our rejections and warnings. Hence my difficulty with Dessaur, *De achtste scheppingsdag*. For criticism, see Bloemers, "De nachtmerrie van Dessaur."

and experimentation. If it is impossible without the destruction of embryonic human life, then we must reject germline engineering.

Finally, as stewards, human beings are called to *preserve* what God has created. Plans for refining human beings have nothing to do with preserving God's creation. When we pursue genetic eugenics we are trying to do better than God. That is no longer an exercise of our stewardship in which we cooperate with God. Instead, we will be in competition with him[54] by forming human beings after our own image.

It is my conviction that stewardship—characterized by working, healing, protecting, and preserving—is the correct framework that will guide us safely through the labyrinth of genetic engineering.

---

54. Eibach, "Soll der Mensch Schöpfer spielen?," 42. Eibach points to the ancient theological concept of "cooperatio cum Deo," which may not be interpreted in a competitive sense.

Bibliography

Aalders, Gerhard Jean Daniël. *Van huisgemeente tot wereldkerk*. Kampen: Kok, 1977.
Aninga, J. B. "Techniek." In *Grote Winkler Prins Encyclopedie*, 8th ed., vol. 21.
Antébi, Elizabeth, and David Fishlock, editors. *Biotechnology: Strategies for Life*. Cambridge, MA: MIT Press, 1986.
Auer, Alfons. *Umweltethik: Ein theologischer Beitrag zur ökologischen Diskussion*. 2nd ed. Düsseldorf: Patmos, 1985.
Bakker, Jac, et al. *Landbouwbeleid in samenhang*. Barneveld: Publikatie van de Groen van Prinsterer Stichting, 1987.
Barth, Karl. *Church Dogmatics*. Vol. III/1: *The Doctrine of Creation*, part 1. Edited by G. W. Bromiley and T. F. Torrance. Translated by J. W. Edwards, O. Bussey, and Harold Knight. Edinburgh: T. & T. Clark, 1958.
———. *Church Dogmatics*. Vol. III/4: *The Doctrine of Creation*, part 4. Edited by G. W. Bromiley and T. F. Torrance. Translated by A. T. Mackay, T. H. L. Parker, Harold Knight, Henry A. Kennedy, and John Marks. Edinburgh: T. & T. Clark, 1961.
Berkhof, Hendrikus. "God in Nature and History." In *New Directions in Faith and Order*. Geneva: World Council of Churches, 1968.
———. *Man in Transit*. Wheaton, IL: Key Publishers, 1971.
Bezitten of bezeten zijn. In *Kerkinformatie*, February, 1975.
Biotechnologie, uitdaging of bedreiging? Supplement to *Boer en Tuinder*, October 28, 1988.
Bibo, H. "Wat is er aan de hand met de aarde?" In *Als de schepping zucht*, edited by Tini Brugge, 23–36. Kampen: Kok, 1987.
Bloemers, H. P. J. "De nachtmerrie van Dessau: Zin en onzin van genetische manipulatie." *NRC/Handelsblad*, November 8, 1988.
Blok, M. J. C. "Met reikhalzend verlangen." *De Reformatie* 57 (1981–82).
Bockmühl, Klaus. *Umweltschutz—Lebenserhaltung*. Giessen: Brunnen-Verlag, 1975.
Boersema, J. J., et al. *Basisboek Milieukunde*. Meppel: Boom, 1984.
Boulding, Kenneth E. "The Economics of the Coming Spaceship Earth." In *Environmental Quality in a Growing Economy: Essays from the Sixth RFF*

Bibliography

Forum, edited by Henry Jarrett, 3-14. Baltimore: John Hopkins University Press, 1966.
Bouma, Hans. *We leven op een slagveld*. Kampen: Kok, 1978.
Carson, Ruth. *Silent Spring*. Boston: Houghton Mifflin, 1962.
Catechism of the Catholic Church. Libreria Editrice Vaticana, 1993. Online: http://www.vatican.va/archive/ccc_css/archive/catechism/p3s2c2a7.htm.
Clark, Jerry L. "Thus Spoke Chief Seattle: The Story of an Undocumented Speech." *Prologue Magazine* 18/1 (1985). Online: http://www.archives.gov/publications/prologue/1985/spring/chief-seattle.html.
Commissie van de Gezondheidsraad. *Advies inzake kunstmatige voortplanting*. Den Haag: Gezondheidsraad, 1986.
Commission of Inquiry of the German Bundestag. *Chancen und Risken des Gentechnologie*. Bonn: 1987.
Cox, Harvey. *The Secular City: Secularization and Urbanization in Theological Perspective*. New York: Macmillan, 1965.
Crul, Marcel, et al. *Natuurlijke hulpbronnen: van verbruik naar beheer*. Meppel: Boom, 1986.
De Vries, A. Ph. "Genentechnologie en bevruchtingsbiologie," *Nederlands Dagblad*, January 4, 1986.
Dekker, Gerard, et al., editors. *Werken: zin of geen zin*. Baarn: Ten Have, 1986.
Derr, Thomas Sieger. *Ecologie en bevrijding*. Baarn: Het Wereldvenster, 1975.
Descartes, René. *Discours de la Méthode*. Edited by Étienne Gilson. Paris: Librairie Philosophique J. Vrin, 1967.
Dessaur, Catharina Irma. *De achtste scheppingsdag*. Arnhem: Gouda Quint, 1988.
Douma, Jochem. *Abortus*. 5th ed. Kampen: Van den Berg, 1984.
———. *Aids—meer dan een ziekte*. Kampen: Van den Berg, 1987.
———. *Algemene Genade*. Goes: Oosterbaan & Le Cointre, 1966.
———. *Gewapende vrede*. 4th ed. Kampen: Van den Berg, 1988.
———. *Rondom de dood*. Kampen: Van den Berg, 1984.
——— *Seksualiteit en huwelijk*. Kampen: Van den Berg, 1993.
———. *The Ten Commandments: Manual for the Christian Life*. Translated by Nelson D. Kloosterman. 1985. Phillipsburg: Presbyterian and Reformed, 1996.
———. *Responsible Conduct: Principles of Christian Ethics*. Translated by Nelson D. Kloosterman. 1983. Phillipsburg: Presbyterian and Reformed, 2003.
———. *Vrede in de maatschappij*. Kampen: Kok, 1985.
Dunstan, Gordon Reginald. "The Human Embryo in the Western Moral Tradition." In *The Status of the Human Embryo*, edited by Gordon Reginald Dunstan and Mary J. Seller, 39-73. London: King Edward's Hospital Fund for London, 1988.
Eibach, Ulrich. *Experimentierfeld: Werdendes Leben*. Göttingen: Vandenhoeck & Ruprecht, 1983.
———. *Gentechnik: Der Griff nach dem Leben*. 2nd ed. Wuppertal: Brockhaus, 1988.

Bibliography

———. "Soll der Mensch Schöpfer spielen?" In *Ethische en maatschappelijke aspecten van de moderne Gentechnologie*. Ede: Prof. D. G. A. Lindeboom Instituut, 1988.
Ellul, Jacques. *The Technological Society*. New York: Knopf, 1964.
Friedrich, Gerhard. Ökologie und Bibel. Stuttgart: W. Kohlhammer, 1982.
Galjaard, Hans, et al. *Voorkomen is beter dan niet genezen*. 3rd ed. Nijkerk: Callenbach, 1978.
Grässer, Erich. "Kai èn meta toon thèrioon (Mark 1, 13b)." In *Studien zum Text und zur Ethik des neuen Testaments*, edited by Wolfgang Schrage. Berlin: de Gruyter, 1986.
Greijdanus, Seakle. *De Openbaring des Heeren aan Johannes*. Amsterdam: H. A. Van Bottenburg, 1925.
Grote Winkler Prins Encyclopedie. E8th ed. Amsterdam: Elsevier, 1984.
Hamer, H. E., and R. Neu. In *Umwelt*, edited by Michael Klöcker and Udo Tworushcka. Ethik der Religionen—Lehre und Leben 5. Göttingen: Vandenhoeck & Ruprecht, 1986.
Häring, Bernhard. *Manipulation: Ethical Boundaries of Medical, Behavioural and Genetic Manipulation*. Slough: St. Paul, 1975.
Hermans, Hans, et al., editors. *DNA-onderzoek. Pion in het spel van goed en kwaad?* Amsterdam: Voorlopige dienst wetenschapsvoorliching, 1978.
Het Tussenrapport: Basis voor brede maatschappelijke discussie. Den Haag: Stuurgroep Maatschappelijke Discussie Energiebeleid, 1983.
Houtman, Cornelis. *Wereld en tegenwereld*. Baarn: Ten Have, 1982.
Hübner, Jürgen. *Die neue Verantwortung für das Leben*. München: Kaiser, 1986.
Huijgen, J. "Creativiteit nodig bij uitwerking nieuwe oriëntatie in de landbouw." *Nederlands Dagblad*, March 11, 1988.
Humber, James M., and Robert F. Almeder, editors. *Biomedical Ethics Reviews 1983*. Clifton, NJ: Humana, 1983.
Huppes, Tjerk. *Een niew ambachtelijk elan*. Leiden: Stenfert Kroese, 1985.
Jacobi, Claus. *De menselijke springvloed*. Baarn: Meulenhoff, 1970.
Jager, Okke. *Schrale troost in magere jaren*. Baarn: Ten Have, 1976.
Jeuken, Marie. *Ethiek*. Assen: Van Gorcum, 1977.
Jochemsen, H., et al. *De status van het menselijke embryo*. Ede: Prof. Dr. G. A. Lindeboom Instituut, 1988.
Jones, D. Gareth. *Manufacturing Humans*. Leicester: Inter-Varsity Press, 1987.
Kayzer, Wim, et al. *Beter dan God*. Broadcast by the Vrijzinnig Protestantse Radio Omroep, March 1987. Hilversum: Video Nederlands Filminstituut, 1989. DVD.
Kayzer, Wim. *Beter dan God: een jaar later*. Broadcast by the Vrijzinnig Protestantse Radio Omroep, March 1988. Hilversum: Vrijzinnig Protestantse Radio Omroep, 1988. VHS.
Keil, Carl Friedrich. *Biblical Commentary on the Old Testament*. Vol. 1: *The Pentateuch*. Grand Rapids: Eerdmans, 1951.
Klapwijk, Abraham. "Elimination of nitrogen from wastewater through denitrification." PhD diss., Wageningen University and Research Centre, 1978.

Bibliography

Klapwijk, J. "Christus en natuur." Unpublished paper. Kampen Theological University, 1987.
Klöcker, Michael, and Udo Tworushcka. *Umwelt*. Ethik der Religionen—Lehre und Leben 5. Göttingen: Vandenhoeck & Ruprecht, 1986.
Kluxen, W. "Fortpflanzungstechnologien und Menschenwürde." *Allgemeine Zeitschrift für Philosphie* 11 (1986).
Korthals, F., et al. "Artificial Insemination and Surrogacy." *Handelingen Tweede Kamer* 20 706 (1987-88).
Kuyper, Abraham. *Common Grace*. 1.1: Noah-Adam. 1.2: Temptation-Babel. 1.3: Abraham-Parousia. Edited by Jordan J. Ballor and Stephen J. Grabill. Translated by Nelson D. Kloosterman and Ed M. van der Maas. Grand Rapids: Christian's Library, 2013-14.
———. *De Gemeene Gratie*. Vols. 2-3. Amsterdam: Höveker & Wormser, 1903-4.
———. *Pro Rege I-III*. Kampen: Kok, 1911-12.
Liedke, Gerhard. *Im Bauch des Fisches*. Stuttgart: Kreuz-Verlag, 1979.
Long, Thomas A. "Infanticide for Handicapped Infants, Sometimes It's a Metaphysical Dispute." *Journal of Medical Ethics* 14 (1988).
Mahoney, John. *Bioethics and Belief*. London: Sheed & Ward, 1984.
Manenschijn, Gerrit. *Geplunderde aarde, getergde hemel*. Kampen: Ten Have, 1988.
Maximilianus, P. *St. Francis' Zonnelied*. 's Hertogenbosch: 1924.
Meadows, Donella H., et al. *Limits to Growth*. New York: New American Library, 1972.
Meijer, F. "Biotechnologie bij het rund." Unpublished paper, 1988.
Mesarovic, Mihaljo, and Eduard Pestel. *Mankind at the Turning Point: The Second Report to the Club of Rome*. New York: E. P. Dutton, 1974.
Mink, Erik. *Kernenergie in opspraak*. Kampen: Kok, 1981.
Moltmann, Jürgen, *God in Creation: A New Theology of Creation and the Spirit of God*. Translated by Margaret Kohl. The Gifford Lectures, 1984-85. San Francisco: Harper & Row, 1985.
Nijkamp, Peter. *Naar een maatschappij zonder toekomst?* Groningen: Uitgeverij De Vuurbaak, 1976.
Nijkamp, Peter, and Jochem Douma. *Het gelaat van de aarde*. Groningen: De Vuurbaak, 1974.
Pestel, Eduard. *Beyond the Limits to Growth: A Report to the Club of Rome*. New York: Universe Books, 1989.
Phillips, Anthony. *Deuteronomy*. Cambridge: Cambridge University Press, 1973.
Rachels, James. *The End of Life: Euthanasia and Morality*. Oxford: Oxford University Press, 1986.
Rahner, Karl. *Schriften zur Theologie VIII*. Ensiedeln: Benziger, 1967.
Reijnders, Lucas. *Pleidooi voor een duurzame relatie met het milieu*. Amsterdam: Van Gennep, 1984.

Bibliography

Reiter, Johannes, and Ursel Theile. *Genetik und Moral*. Mainz: Matthias-Grünewald-Verlag, 1985.
Schaeffer, Francis A.. *Pollution and the Death of Man: The Christian View of Ecology*. Wheaton, IL: Tyndale House, 1970.
Schellekens, Hubertus, and Robert Paul Willem Visser. *De genetische manipulatie*. Amsterdam: Meulenhoff Informatief, 1987.
Schilder, Klaas. *Christ and Culture*. Translated by G. van Rongen and W. Helder. Winnipeg: Premier Printing, 1977.
Schilpzand, R. In *Boerderij*, February 17, 1987.
Schlink, Edmund. In *New Directions in Faith and Order, Bristol 1967*. Geneva: World Council of Churches, 1968.
Schroten, Egbert. *In statu nascendi: De beschermwaardigheid van het menselijk embryo vanuit het gezichtspunt van de christelijke ethiek*. Utrecht: Wever, 1988.
Schumacher, Ernst Friedrich. *Small Is Beautiful: Economics as If People Mattered*. New York: Harper & Row, 1973.
Schuurman, Egbert. *Crisis in de landbouw*. Wageningen: Stichting voor Reformatorische Wijsbegeerte, 1987.
———. *Technology and the Future: A Philosophical Challenge*. Translated by Herbert Donald Morton. Toronto: Wedge Publishing, 1980.
———. "Wijsgerig-ethische en maatschappelijke aspecten van de moderne Gentechnologie." In *Ethische en maatschappelijke Aspecten van de moderne Gentechnologie*. Ede: Prof. Dr. G. A. Lindeboom Instituut, 1988.
Seldenrijk, Ruth. *Genetische technieken en christelijke ethiek: Sleutelen aan erfelijkheid in gezondheidszorg en landbouw*. Houten: Den Hertog, 1988.
Seller, Mary J., and Elliot Philipp. "Reasons for Wishing to Perform Research on Human Embryos." In *The Status of the Human Embryo*, edited by Gordon Reginald Dunstan and Mary J. Seller, 22–32. London: King Edward's Hospital Fund for London, 1988.
Singer, Peter. *Practical Ethics*. Cambridge: Cambridge University Press, 1979.
———, and Deane Wells. *The Reproduction Revolution: New Ways of Making Babies*. Oxford: Oxford University Press, 1984.
Slijper, Everhard Johan. *Het lot der mensheid*. Groningen: Wolters, 1952.
Smelik, Evert Louis. *De ethiek in de verkondiging*. 3rd ed. Nijkerk: Callenbach, 1967.
Spinoza, Benedictus de. *Ethics*. Edited and translated by G. H. R. Parkinson. Oxford: Oxford University Press, 2000.
Steck, Odil Hannes. *Welt und Umwelt*. Stuttgart: W. Kohlhammer, 1978.
Steering Committee of the Council of Churches in the Netherlands. *Een verbond voor het leven*. Amersfoort: De Horstink, 1988.
Tinbergen, Jan. *Kunnen wij de aarde beheren?* Kampen: Kok Agora, 1987.
Toffler, Alvin. *The Third Wave*. New York: Bantam, 1980.
Tworushcka, M. In *Umwelt*, edited by Michael Klöcker and Udo Tworushcka. Ethik der Religionen—Lehre und Leben 5. Göttingen: Vandenhoeck & Ruprecht, 1986.

Bibliography

Van Bruggen, Gerbrandina. *Hulpverlening in gebroken situaties: stoppen of doorgaan.* Amersfoort: Vereniging ter Bescherming van het Ongeboren Kind, 1987.

Van den Berg, Hendrik, et al. *Bouwen en bewaren.* Groningen: Groen van Prinsterer Stichting, 1974.

Van den Daele, Wolfgang. *Mensch nach Mass?* München: Beck, 1985.

Van der Poll, Evert, and Janna Stapert. *Als het water bitter is.* Sliedrecht: Merweboek, 1988.

Van Dijk, A. In *Umwelt,* edited by Michael Klöcker and Udo Tworushcka. Ethik der Religionen—Lehre und Leben 5. Göttingen: Vandenhoeck & Ruprecht, 1986.

Van Dijk, Gert, and J. Huygen. "Ethiek en de moderne veehouderij." *Zichtkaternen* 21 (1984).

Van Dijke, J. "Manipuleren met de levende schepping." In *Over de schepping gesproken,* edited by J. Leune et al. Woerden: Jeugdbond Gereformeerde Gemeenten, 1987.

Van Leeuwen, Arend Theodoor. *Christianity in World History.* London: Edinburgh House, 1964.

Van Leeuwen, J. F. M. In *Boerderij/Livestock Operation,* September 6, 1988.

Van Loon, Antonius Johannes. *Kernenergie: voor of tegen?* Utrecht: Spectrum, 1981.

Van Steenis, Cornelis Gijsbert Gerrit. *Homo destruens.* Amsterdam: Noordhoff-Kolff, 1954.

Velema, Willem Hendrik. *Ethiek en pelgrimage.* 2nd ed. Amsterdam: Bolland, 1974.

Verheij, Sigismund. "Franciscus, een mens of aarde." In *Als de schepping zucht,* edited by Tini Brugge. Kampen: Kok, 1987.

―――, et al., editors. *Franciscus van Assisi: Keuze uit artikelen verschenen in het jubileumjaar 1976.*Utrecht: Werkgroep K. 750, 1978.

Von Rad, Gerhard. *Deuteronomy.* Translated by Dorothea Barton. Old Testament Library. Philadelphia: Westminster, 1966.

Von Weiszäcker, Carl Friedrich. *De tijd dringt: Pleidooi voor een vredesconcilie van alle christenen in de wereld over gerechtigheid, vrede en het behoud van de schepping.* Baarn: Ten Have, 1987.

Voous, Karel Hendrik. *Natuur, milieu en mens.* Kampen: Kok, 1970.

Warnock, Mary. *A Question of Life.* Oxford: Blackwell, 1985.

Westermann, Claus. *Genesis 1-11.* Translated by John J. Scullion. Minneapolis: Augsburg, 1984.

White, Lynn Jr. "The Historical Roots of our Ecological Crisis." *Science* 155 (1967) 1203-7.

Wiskerke, Nicolaas. "Meeleven met de schlepping." In *Als de schepping zucht,* edited by Tini Brugge, 104-20. Kampen: Kok, 1987.

Witkam, W. G. M., et al. *In-vitro-fertilisatie.* Ede: Prof. Dr. G. A. Lindeboom Instituut, 1988.

Scripture Index

OLD TESTAMENT

Genesis

1	29, 30, 31, 40, 41, 42, 43, 46
1–11	34
1:11–12	27, 31, 34, 80
1:21	31, 34, 80
1:24–25	31, 34, 80
1:26–27	40
1:26–30	29–30
1:28	7, 39, 45
1:29	27
1:31	30
2	31
2:2–3	35
2:9	27
2:15	31, 38, 88
2:16	27
2:17	119
2:19–20	30
3	29, 31
3:1–15	31
3:17–19	26
3:22	27
4:21–22	22
6:19–20	31
9	30
9:2–3	30
9:4	32, 62
9:9–10	31

Exodus

20:10	28
22:19	80
23:10–11	28
23:11	32

Leviticus

1–7	29
11:44	46
11:46–47	29
17:10–14	62
18:23	80
19:19	80
20:15–16	80
25:1–12	28
25:7	32
25:43	30
25:46	30

Numbers

33:55	26

Scripture Index

Deuteronomy

5:14	28
20:19–20	27
22:6–7	28
22:9	80
22:11	80
25:4	28, 63

Joshua

23:13	26

Judges

9:8–15	26
14:5–6	27

1 Samuel

17:34–37	27

1 Kings

4:33	32
5:6–18	27
5:7	19
13–18	19

2 Kings

14:9	26
19:23	27

2 Chronicles

36:21	28

Nehemiah

13:18	28

Job

28	41
28:1–11	41
31:40	26
38–39	32
38–41	26
40:4	32

Psalms

8:6	91
8	18, 26, 32, 43
19:1	35n22
22:13	28
22:22	28
102:25–26	50
104	26, 32, 36
105	36

Proverbs

6:6–8	32
12:10	28, 63, 85

Ecclesiastes

1:18	97

Song of Solomon

1:17	27
2:2	26

Isaiah

11:6–9	27, 33
41:18–20	27
51:6	50
55:13	26, 27
56:9–12	28

60:13	26	**NEW TESTAMENT**	
65:22	26		

Jeremiah

Matthew

2:15	28	6:25–30	32
4:7	28	6:26–28	26
12:7–8	28	6:34	96
31:12	27	24:14	44
		24:45–46	39

Ezekiel

Mark

17:23	26	1:13b	33
20:13	28		
28:24	26		
39:17–20	28	**John**	
47:12	27	14:12	18, 22

Hosea

Acts

2:17	27	15:20	62n22

Joel

Romans

3:18	27	1:20	38
		5:12–14	119
Amos		7:20	119
		8:18–23	36
2:9	26	8:19–23	62
9:13–15	27	8:20–23	90

Jonah

1 Timothy

4:11	32	6:7	55
		6:8–10	39

Habakkuk

Hebrews

2:17	27	2:5–9	43

Scripture Index

1 Peter		Revelation	
1:16	46	6	53, 53n6
		13:18	91
2 Peter		22:2	27
3:5–11	43	**APOCRYPHAL LITERATURE**	
			30n10

Subject Index

abolition of slavery, 20n31
abortion
    China's one child policy and, 52
    of embryos with defects, 102–103
    genetic engineering of defects and, 105
    medieval view of, 114
    as solution to population growth, 50–51
abstinence, 54–55
acid rain, 4–5, 56, 81–82
acromegaly, 103
ADA (adenosine deaminase) deficiency, 106n21
Adam and Eve
    blessing on, 43n36
    consequences of sin of, 10, 18, 26, 70, 89, 109, 119
agriculture
    development of, 42
    environmental damage from plowing, 11–12
    genetic engineering, 77, 82–85
    mistreatment of animals, 6–7, 64–65
    political policies toward, 56
    pollution from livestock, 5, 56, 84
    pollution from use of chemicals, 6
air pollution, 5–6, 8

Almeder, Robert F., 86m17
Alzheimer's disease, 60
ammonia, 5
amniocentesis, 101
animals
    Biblical respect for, 28, 31–33
    cloning of, 83–85
    cruelty to, 62–63
    experiments on, 63
    as food, 60, 61, 62
    genetic engineering, 74–76, 82–85, 90–91, 92
    God-given abilities, 34
    nature of, 80–85
    as pets, 65
    protection of, 59–61
    respect for, 62–65
    selective breeding, 73–74
    suffering with creation, 62
animism, 7, 12
anthropocentric vision, 33, 34–35, 47, 62
anti-world, 26
Apocryphal literature, 30n10
Aristotle, 12
art, 42
artificial insemination, 83, 84–85
"Artificial insemination" (Korthals), 110n30
Asilomar Conference, 86–87
Aswan Dam, 9
Attfield, R., 37–38n29

133

Subject Index

bacteria
  genetic engineering of, 76–77
  genetic engineering of humans, 87
  patent rights for, 85
  producing chymosin, 92n32
  production of drugs, 103–104
  used for oil clean up, 81–82
Barth, Karl, 62
bestiality, 80
Belgic Confession, article 2, 38
Berg, Paul, 87
Berkhof, H., 17, 43n35
Beter dan God (television program), 76n4, 89, 106n20
Beyond the Limits to Growth (Pestel), 44
Bible
  depiction of plants and animals, 27–28
  doctrine on the environment, 25–47
  immutable ethics of, 79–80, 120
  lessons from animals and plants, 32
  objection to destruction of nature, 27
  prohibition of interbreeding, 80
  respect for plants and animals, 28, 31–33
  See also Scripture Index
Biomedical Ethics (Humber and Almeder), 86m17
biotechnology, 60
birth control, 51–52
black-stocking Christians, 21n33
BMD, 66
Bockmühl, Klaus, 39
Boulding, Kenneth E., 3
bovine somatotropin (BST), 82–83
Brown, Louise, 111
Buddhism, 15
bullfighting, 65n30

businesses, medical examination requirements, 98–99

Cain's descendants, 22
Cairo, 51
Calcutta, 51
Calvinism, 7, 13–14, 23, 37
cancer, 95, 97
Canticle of the Sun (Francis of Assisi), 36
capitalism, 7, 13–14, 23
carbon dioxide, 5, 66
care of creation, 39
cars
  accidents, 69
  political policies concerning, 57
  pollution from, 4–5, 6, 57
Carson, Rachel, 6
Catechism of the Catholic Church, 63n26
cattle, genetic engineering of, 82–83
centralization, 49–50, 69, 100–101
CFCs (chlorofluorocarbons), 5
Chakrabarty, Ananda, 77
Chernobyl nuclear accident, 6, 66, 67, 69
chimeras, 90–91, 118
China
  on child policy, 52
  inventions of, 20
  pets in ancient times, 65
chlorofluorocarbons (CFCs), 5, 55
chorionic villus sampling, 101
Christ and Culture (Schilder), 39–41
Christianity
  liberation through, 18
  relationship to technology, 17–19, 20–24
  relationship to the environment, 11–14
  tradition of subduing the Earth, 7
chromosomes, 75. *See also* DNA (deoxyribonucleic acid);

Subject Index

genes; genetic engineering of animals and plants; genetic engineering of humans; selective breeding
chymosin, 92
circus, 65
climate change, 5–6
cloning
 of animals, 83–85
 of humans, 117
Club of Rome, 1–2
coagulation factor, 103
coal. *See* fossil fuels
Cohen, Stanley N., 87
Coleman, W., 37–38n29
Commission of Inquiry of the German Bundestag, 99n8, 108n27
computer
 centralizing and decentralizing tendencies, 49n2
 danger of, 69
 genetic databases on, 100–101
conflict regulation, 30n9
contentment, 39
cooperatio cum Deo, 121n54
cosmetic experimentation, 63
cosmocentric vision, 33–34, 47
cowboy economy, 3, 53
creation
 nature of, 80–85
 stewardship of, 36–38
 See also animals; plants; stewards
Crick, Francis H. C., 74, 75
Crisis (Schuurman), 89n26
crucifixion of ego, 46, 47, 54
cultural mandate
 in context of Christ's redemptive work, 45
 interpretation of, 47
 meaning of, 39–44
 pilgrimage and, 44–47, 54
cystic fibrosis, 94n2

data banks, 100–101

David (king of Israel), 27
DDT, 6
death, 119
decentralization, 49–50, 56
deforestation, 4, 6
Delta Plan, 67
democracy, 13
De Openbaring (Greijdanus), 91n31
deoxyribonucleic acid (DNA). *See* DNA (deoxyribonucleic acid)
Descartes, René, 7, 12, 21
desertification, 4, 8–9
De status van het menselijke embryo (Jochemsen), 113n38
diabetes, 94n2, 104
Die neue Verantwortung (Hübner), 88n22
disengagement, 54
DNA (deoxyribonucleic acid)
 as evidence in criminal cases, 77
 recombinant DNA technology, 75–76, 103–104
 research on, 88–89
 structure of, 74–75
 See also chromosomes; genes; genetic engineering of animals and plants; genetic engineering of humans; selective breeding
DNA fingerprinting, 100
doctors' responsibility, 97–98n7
doctrine, 14–16
Down syndrome, 102, 105
droughts, 4
drugs, 103
Duchenne muscular dystrophy, 105
dwarfism, 103

Eastern Orthodoxy, 23n38
echography, 101
ecology, 3–4
economic factors
 involved in solutions to damage, 48–49, 58

135

Subject Index

economic factors, *continued*
   limits of growth, 1
   of nuclear power, 67
   population control and, 52
   as source of environmental
      damage, 7
ecosystem, 3–4
Edwards, Robert, 111
Egypt, 19, 20n31
Eibach, Ulrich, 86n18, 89n26, 121n54
election, doctrine of, 13–14
Ellul, Jacques, 20, 23n38
embryo cleavage, 84n13
embryo transfer, 83, 85
England, 58
Enlightenment, 7, 21, 22
environment
   Christianity's view of, 11–14
   defined, 3–4
environmental damage
   deforestation, 4
   desertification, 4
   guilty parties, 7–24
   limits of population and economic growth, 1
   solutions to, 46, 48–71, 76–77
environmental pollution
   bacteria used for clean up, 76–77, 81–82
   limits of, 1–3
   nullification of profitability, 38
   sources of, 3–7
Escherichia coli, 103
ethics
   definition of, 79
   of gene therapy, 105–106, 120
   on genetic engineering, 78–80, 88n22
Ethics and Pilgrimage (Velema), 45
eugenics, 115–119, 121
euthanasia, 50–51
eutrophication, 5
evangelism, 44–47, 54

evolution, 73–74
experimentation, 86–88
exploitation, 38
extinction, 4, 6, 8

fall of mankind, 10, 18, 26, 70, 89, 109, 119
Faraday, Michael, 69
farming industry. *See* agriculture
fasting, 54
fishing, 6
floods
   Noahic covenant, 31–32
   as result of deforestation, 4, 5, 8
"Fortpflanzungstechnologien" (Kluxen), 112n36
fossil fuels
   availability of, 67
   nuclear energy vs., 66
   nuclear fusion and, 68
   pollution from burning process, 5, 6, 66
Fourth Commandment, 32
France, 58
Francis of Assisi, 36

Galjaard, Hans, 89n25, 102n14
Garden of Eden, 31. *See also* Paradise
gas. *See* fossil fuels
gene passport. *See* genetic diagnosis (gene mapping)
genes
   identification of disease-causing genes, 94
   mapping of, 75
   role of, 95
   structure and number of, 75, 94
   See also chromosomes; DNA (deoxyribonucleic acid); genetic engineering of animals and plants; genetic engineering of humans; selective breeding
genetic determinism, 95

Subject Index

genetic diagnosis (gene mapping), 94–101
  genetic databases, 100–101
  insurance applications, 99–100
  medical examinations, 98–99
  right to remain uninformed, 95–98
genetic discrimination, 101
genetic engineering of animals and plants, 72–92
  acceptable applications, 76–77
  barriers overcome, 74–76
  danger of experimentation, 86–88
  ethics and, 78–80
  goals of, 78, 87–88
  moratorium on, 86–87
  need for definition of boundaries, 42
  patent rights for living organisms, 77, 85–86
  people as stewards and, 88–92
genetic engineering of humans, 93–121
  conclusion, 120–121
  genetic diagnosis (gene mapping), 94–101
  genetic eugenics, 115–119
  genetic selection, 101–103
  genetic therapy, 103–115
  stewardship and, 118–119, 120–121
genetic erosion, 84
genetic eugenics, 115–119, 121
genetic selection, 101–103
genetic therapy, 103–115
  experimenting with pre-embryos, 110–114
  germline engineering, 107–110, 114, 116, 121
  somatic gene therapy, 103–107, 108, 120
  theologians paving the way, 114–115

genotype, 95n4
Gentechnik (Eibach), 86n18, 89n26, 106n23
Germany, 58–59
germline engineering, 107–110, 114, 116, 121
Gilson, Étienne, 21n32
God in Creation (Moltmann), 42n33
Graham, Billy, 53n6
Grässer, Erich, 30n10
greed, 39
Greeks, ancient, 19
greenhouse effect, 5
Greens, 81
Greijdanus, Seakle, 91n31
growth hormone, 103–104
guilty parties, 7–24

Harrisburg nuclear accident, 6, 66
health, defined, 116
hemophiliacs, 103
Hinduism, 15
HIV testing, 100
homosexuality, 50–51
Houtman, Cornelis, 26
Hübner, Jürgen, 88n22
human embryos
  experimentation on, 110–114
  genetic engineering of, 105
human genome
  defined, 94n1
  mapping of, 94–101
humanism, 22
humanitarian activities, 51
human laboratory models, 106–107
humans
  as central in Biblical passages, 29
  chromosomes of, 75
  cloning of, 117–118
  dominion over nature, 29–31, 80
  genetic engineering, 93–121
  limitations of, 31
  See also fall of mankind; stewards
Humber, James M., 86m17

137

Subject Index

hunting, 6, 8, 60, 61, 62
Huntington's disease, 95, 96, 105
Huygen, J., 85n15
hybrid goat, 76n3
hybridization, 73–74
hybrid sheep, 76n3
hydroelectric power, 67

Im Bauch des Fisches (Liedke), 30n9
India, 52
Industrial Revolution, 7
  as source of environmental damage, 12
industry
  government regulation of, 56, 57
  as source of environmental damage, 5, 38, 56
infanticide, 52
"Infanticide" (Long), 113n38
infant screening, 98
insulin, 104
insurance, 99–100
interferon, 104
international cooperation, 58–59
Islam, 14, 20
Israelites
  abstention from blood of animals, 62
  appreciation of nature, 26, 27
  prohibition of interbreeding, 80
  ritual slaughter, 64
  struggle with the environment, 26
  technology of, 20n31
  See also Jews

Jacobi, Claus, 11, 25, 46
Jager, Okke, 37–38n29
Jakarta, 51
Japan, 15, 16, 23n37
Java, 52
Jerusalem Temple, 19
Jesus Christ
  on anxiety about the future, 96, 97
  dominion over nature, 18–19, 43, 45
  gift of eternal life, 119
  redemptive work of, 54
  reformative task of, 41
  temptation in the wilderness, 33
Jews, 64. *See also* Israelites
Job, 32
Jochemsen, H., 113n38
Jones, D. Gareth, 112n37
Jubilee years, 28, 32

"Kai èn meta toon thèrioon" (Grässer), 30n10
Kluxen, W., 112n36
Korthals, F., 110n30
kosher foods, 29, 62, 64
Kuyper, Abraham
  on cultural task of man, 40, 42, 47
  on human development, 22
  on humans' power over nature, 17–19

Leene, H., 43n36
Leeuwen, A. Th. van, 17
Lejeune, J., 111
Libya, 8
Liedke, Gerhard, 30n9
life, 14–16
lifeboat ethic, 50–51
life span, 119
Limits to Growth, The (Club of Rome), 1–3

magic, 20
malaria, 106n20
Manenschijn, Gerrit, 37–38n29
Manilla, 51
Man in Transit (Berkhof), 17
Manipulatie (Schellekens and Visser), 94n2, 106n21

Subject Index

Manufacturing Humans (Jones), 112n37
meat-eating, 60, 61, 62
medical examinations, 98–99
medicines, 76, 103
Meijer, F., 85n15
Mexico City, 51
Miescher, Friedrich, 74
mining, 8, 42, 67
Moltmann, Jürgen, 42n33
Mumbai, 51
Muslims, 14, 20, 64
mutations, 73

Native American religion, 15
Nature (magazine), 75–76
nature-nurture interaction, 95
negative eugenics, 115
Netherlands
    case involving DNA testing, 77
    Delta Plan, 67
    fossil fuel supplies, 67
    pollution of, 58
New Directions in Faith and Order (Schlink), 43n35
new heaven and earth, 50
Nineveh, 32
Noahic covenant, 30, 31–32
Noah's arc, 31
nuclear energy
    discovery of, 72
    fission vs. fusion, 67–68
    need for definition of boundaries, 42
    nuclear power plant accidents, 6, 66, 67, 69
    as potential hazard to environment, 66–67, 72
    as solution to energy needs, 67–68, 69–70
nuclear weapons, 42, 68, 72
number 666, 91

oceans, 5, 6

oil. *See* fossil fuels
one-world government, 58
organic growth and development, 2–3
overpopulation, 7
ozone layer, 5

paganism, 12, 20, 26
pantheism, 15–16
Paradise
    relationship between man and nature in, 30–31, 33
    trees of, 27
    as unfinished world, 40
    See also Garden of Eden
patent rights, 77, 85–86
peace, 54n8
persistent pesticides, 6
Pestel, Eduard, 2–3, 44, 57
pesticides, 6
pets, 65
phenotype, 95n4
Philipp, Elliot, 110n31
philosophers, 7
pigs, 82–83
pilgrimage, 44–47, 54
PKU testing, 98
plants
    Biblical respect for, 28, 31–33
    genetic engineering, 77, 81, 85–86, 91–92
    genetic erosion, 84
    God-given abilities, 34
    hybridization and selective breeding, 73–74
    nature of, 80–85
plow, 11–12, 19, 20
pluripotent cells, 107n25
political policies
    concerning cloning, 118
    concerning gene mapping, 96, 99
    as solutions to environmental damage, 56–59

Subject Index

Pollution and the Death of Man
    (Schaeffer), 21n33
population
    improper methods of control,
        50–51
    limits of growth, 1–2
    projected growth of, 51
porcine somatotropin (PST), 82–83
positive eugenics, 115, 119
pre-embryo experimentation,
    110–115
prenatal intervention, 102
prenatal research, 101
profitability, 37–38
pro-life, defined, 103
Protestantism, 13–14

Rachels, James, 60–61
radioactivity
    in the atmosphere, 6
    genetic changes caused by, 73
    nuclear power plant accidents, 6,
        66, 67, 69
rationalism, 22
rdh (dominion), 30n9
recombinant DNA technology, 75
recombinant embryonic cells, 76n3
recycling, 55, 56
reforestation, 4, 8
Reformation, 21
relief actions, 51
Renaissance, 22
rennin, 92
"Research on Human Embryos"
    (Seller and Philipp), 110n31
right to remain uninformed, 95–98,
    99n8
ritual slaughter, 64
rodeos, 65n30
Roman Empire
    environmental pollution in, 8, 16
    pets in, 65
    use of technology, 19–20, 22
Russia, 23n38

Sabbath rest, 28, 32, 35
Sakharov, Andrei, 68
Samson, 27
Scandinavia, 58
Schaeffer, Francis A., 21n33
Schellekens, Hubertus, 94n2,
    106n20, 106n21
Schilder, K., 39–41, 42–43, 80n8
Schlink, Edmund, 43n35
Schroten, Egbert, 114n42
Schumacher, Ernst Friedrich, 49n1
Schuurman, Egbert, 89n24, 89n26
Schweitzer, Albert, 34
science, 13
Seattle (Indian chief), 15
selective breeding, 73–74, 83
self-denial, 46, 47, 54–55
Seller, Mary J., 110n31
serfdom, 20
Shintoism, 15, 16
sickle-cell anemia, 106n20
Silent Spring (Carson), 6
sin, consequences of, 10, 18, 26, 70,
    89, 109, 119
Singapore, 56–57
Singer, Peter, 59–60, 61
slaughterhouses, 62
slavery, abolition of, 20n31
Small is Beautiful (Schumacher),
    49n1
social invalidation, 101
solar power, 67
"Soll der Mensch Schöpfer spielen?"
    (Eibach), 121n54
Solomon (king of Israel), 19, 32
solutions
    attitude for workers, 49–50
    complexity of situation and, 46,
        48–49
    genetic engineered bacteria,
        76–77
    impermissible methods, 50–52
    nuclear energy as, 66–68
    personal attitudes toward, 53–55

political policies, 56-59
respect for animals, 59-65
using common sense, 69-71
somatic gene therapy, 104-107, 109n28, 118, 120
somatotropin, 104
speciesism, 59-60, 61
Spinoza, Benedictus de, 62
Steck, Odil Hannes, 30n11
Steering Committee of the Council of Churches in the Netherlands, 54n8
Steptoe, Patrick, 111
stewards
  conduct of, 39
  as cultural mandate, 39-41
  genetic engineering and, 88-92, 118-119, 120-121
  proper attitude toward creation, 36-38
sulfur dioxide, 5, 66
supermouse, 75-76

Taoism, 11
technology
  consequences of, 69-70
  as general phenomenon, 19-24
  relationship to Christianity, 17-19, 20-24
  as source of environmental damage, 7
  unity with science, 13
temperatures, 5
tenant, 37
Ten Commandments, The (Douma), 85n15
theocentric vision

evaluation of, 41-44
stewardship of man, 37-41
view of creation and humans, 34-35, 47
theologians, 114-115
Third Wave, The (Toffler), 49n2
Time (magazine), 94, 100
Toffler, Alvin, 49n2
tourism, 4

urbanization, 51, 56
U.S. Supreme Court, 86

vaccines, 104
Van Dijk, Gert, 85n15
Velema, W. H., 45
Visser, Robert Paul Willem, 94n2, 106n21
in vitro fertilization, 111
vivisection, 60, 62, 63
Voorkomen (Galjaard), 102n14

waste disposal, 55, 56
watermills, 20
water pollution, 5, 6, 8, 56
Watson, James D., 74, 75
Watt, James, 69
Weber, Max, 13-14, 23
Welt und Umwelt (Steck), 30n11
Western Europe, 52
White, Lynn, 11-13, 17, 20
wholesomeness, 54n8
wind energy, 67, 69
windmills, 20
World Health Organization, 116

zoos, 65

www.ingramcontent.com/pod-product-compliance
Lightning Source LLC
Chambersburg PA
CBHW072148160426
43197CB00012B/2298